WARWICK INTERNATIONAL LIMITED
MOSTYN, HOLYWELL, CLWYD CH8 9HE

Tel: (0745) 560651
Telex : 61640
Fax : 560190

DATA ANALYSIS IN THE CHEMICAL INDUSTRY
Volume 1: Basic Techniques

ELLIS HORWOOD SERIES IN INORGANIC CHEMISTRY

Series Editor: J. BURGESS, Department of Chemistry, University of Leicester

Inorganic chemistry is a flourishing discipline in its own right and also plays a key role in many areas of organometallic, physical, biological, and industrial chemistry. This series is developed to reflect these various aspects of the subject from all levels of undergraduate teaching into the upper bracket of research.

DATA ANALYSIS IN THE CHEMICAL INDUSTRY
Volume 1: Basic Techniques

R. CAULCUTT, B.Sc.
BP Chemicals Lecturer
The Management Centre, University of Bradford

ELLIS HORWOOD LIMITED
Publishers · Chichester

Halsted Press: a division of
JOHN WILEY & SONS
New York · Chichester · Brisbane · Toronto

First published in 1989 by
ELLIS HORWOOD LIMITED
Market Cross House, Cooper Street,
Chichester, West Sussex, PO19 1EB, England
The publisher's colophon is reproduced from James Gillison's drawing of the ancient Market Cross, Chichester.

Distributors:

Australia and New Zealand:
JACARANDA WILEY LIMITED
GPO Box 859, Brisbane, Queensland 4001, Australia

Canada:
JOHN WILEY & SONS CANADA LIMITED
22 Worcester Road, Rexdale, Ontario, Canada

Europe and Africa:
JOHN WILEY & SONS LIMITED
Baffins Lane, Chichester, West Sussex, England

North and South America and the rest of the world:
Halsted Press: a division of
JOHN WILEY & SONS
605 Third Avenue, New York, NY 10158, USA

South-East Asia
JOHN WILEY & SONS (SEA) PTE LIMITED
37 Jalan Pemimpin # 05–04
Block B, Union Industrial Building, Singapore 2057

Indian Subcontinent
WILEY EASTERN LIMITED
4835/24 Ansari Road
Daryaganj, New Delhi 110002, India

© **1989 R. Caulcutt/Ellis Horwood Limited**

British Library Cataloguing in Publication Data
Caulcutt, Roland
Data analysis in the chemical industry
Vol. 1: Basic techniques.
1. Chemistry. Applications of statistical mathematics
I. Title
540'.1'5195

Library of Congress data available.

ISBN 0–7458–0727–5 (Ellis Horwood Limited)
ISBN 0–470–21492–9 (Halsted Press)

Typeset in Times by Ellis Horwood Limited
Printed in Great Britain by Hartnolls, Bodmin

Table of contents

Preface

For many years I have attempted to plough two furrows. On the one hand I have presented short courses in applied statistics for the chemical industry whilst on the other hand I have studied and taught educational psychology. This book is based on both experiences. It is concerned with the statistical analysis of data and it is written in a format which is based on sound psychological principles.

> In my opinion many people will learn more by studying this book than they would learn by attending a four- or five-day course.

No self-study text can give you *everything* that you could obtain from a good course. For example, you cannot get the stimulus of social interaction which is an attractive feature of many residential courses. On the other hand, guided self-study offers you greater opportunity to *think*. Indeed, this book is written in such a way that it compels you to reflect upon, and to apply, what you have read. Furthermore, the thinking occurs frequently as there are short exercises at regular intervals throughout the text. I strongly recommend that you attempt each exercise then study its worked solution immediately afterwards.

The content of this book has been field-tested by hundreds of industrial scientists and technologists and by many classes of chemistry students at Huddersfield Polytechnic. I would like to record my thanks to all those who suggested improvements. The incorporation of these suggestions has, in my opinion, raised the quality of the text to true 'teach yourself' standard. Unfortunately, no amount of quality improvement could produce a book which was able to impart instant wisdom to all readers. The learning process is inherently slow. For effective learning to take place, the reader must reflect upon how the new knowledge relates to previous experience and understanding. This reflection is very time-consuming. Therefore it is not a prominent feature of short courses.

Each chapter of this book contains **detailed learning objectives**. These are placed towards the **end** of the chapter, immediately before the self-test. the objectives tell you what you should be able to do **after** you have studied the learning material within the chapter. In contrast, the **aims** appear in the introduction near the **beginning** of the chapter. You will find that the aims are written in more general terms and can be understood without any detailed knowledge of the chapter's content, whereas the objectives are more specific and would appear rather forbidding if they appeared earlier.

Before you start to read this book you may wish to know how long it will take you to master all that it has to offer. Experience suggests that each chapter may require about five hours of undivided attention, though some readers may wish to proceed more slowly. Of course, there is no necessity to devour each chapter in one continuous session. I am sure that many readers will prefer to study for shorter spells, perhaps having a break after each activity. If, at the outset, you know that you cannot spare the thirty-five or more hours needed to study the whole book, you may be pleased to learn that the first three chapters, constitute a well-balanced and comprehensive introductory course. For the reader who has time for only four chapters, I would recommend chapters 1, 2, 3 and 7, in order to maximize your awareness of what can be achieved using simple statistical techniques. On the other hand, any reader who wishes to progress to the more advanced techniques in later volumes of this series, would be well advised to build a solid foundation by carefully studying every chapter of this text.

I would like to thank Mike Hold and Angela James of Huddersfield Polytechnic, for the help they gave me in the writing of this book. Mike showed great patience whilst guiding me through the complexities of desktop publishing. Angela must have had even greater patience to have produced such a good typescript from my misspelt and barely legible original. Any errors that remain are entirely my responsibility.

I hope that you will enjoy studying this book and I predict that the time and effort you devote to the task will prove to be a profitable long-term investment. Should you feel the need for a short practical/tutorial course to supplement your self-study, details of such courses can be obtained from: Roland Caulcutt, The Management Centre, University of Bradford, Emm Lane, Bradford BD9 4JL, Tel.: (0274) 542299.

1

Assessing quality and quantity

1.1 INTRODUCTION

The first chapter in many statistics books is a leisurely preamble in which the reader is shown how to carry out simple calculations and to draw simple diagrams. For many readers this can be a terrible bore. Furthermore, such a chapter can give a distorted impression of statistics, which has so much to offer to the industrial scientist or technologist. The second chapter of the conventional text is often devoted to the foundations of probability theory. This may be more interesting but, like the first chapter, it gives the reader no indication of how statistics can be used. Indeed, there may be several chapters of progressively more difficult theory before the author reveals the usefulness of statistical techniques.

In this text I have adopted a radically different approach. Chapter 1 and all subsequent chapters are devoted to **applications** of statistics. Each statistical technique is applied to a real-life problem, with which the reader can identify. The first of these problems is concerned with the performance of a production process which has been recently modified. Only after we have examined several problems will I attempt to answer the question 'What is statistics?' Chapter 1 has three aims, which you should be aware of at the outset:

(1) I intend that this chapter will give you a fairly gentle introduction to the statistical analysis of data. This will be achieved by focusing on two specific techniques, which are useful in themselves, but which also serve to illustrate the nature of statistical analysis.
(2) I expect that you will acquire confidence in your ability to use these techniques as an aid to the drawing of valid conclusions. A discussion of the assumptions underlying the methods should enable you to identify situations in which they can, and cannot, be used.
(3) It is hoped that you will begin to appreciate the particular perspective adopted by

statisticians towards data analysis. An awareness of this perspective will help you if you need to consult a statistician or to study other statistical texts.

1.2 ASSESSING THE YIELD OF A PROCESS

Higson Industrial Chemists manufacture a range of pigments for use in the textile industry. One particular pigment, digozo blue, is made by a well-established process in a plant which has recently been renovated. During the renovation various modifications were incorporated, one of which made the agitation system fully automatic. Although this programme of work was very successful in reducing the number of operators needed to run the plant, production of digozo blue has not been completely trouble free since the work was completed. Firstly, the expected increase in yield has not been achieved in every batch; secondly, one batch has been found to contain a disturbingly large percentage of a particular impurity. The yield and impurity of the first six batches are given in Table 1.1.

Table 1.1 — Six batches of digozo blue pigment

Batch	Yield (kg)	Impurity (%)
1	690	1.63
2	721	5.64
3	643	1.03
4	741	0.56
5	681	1.66
6	712	1.90

The Plant Manager would like to use the data in Table 1.1 to assess the performance of the process. He also has access to a considerable amount of data gathered before the plant was renovated, but this may now be irrelevant or misleading. He particularly wishes to estimate the yield and impurity that are likely to be found in future batches. Let us concentrate firstly on the yield. What conclusions can Dr Murphy draw about future batches from the data in Table 1.1?

Exercise 1.1

Examine very carefully the six yields in Table 1.1. Using common sense, or intuition, or other means, advise Dr Murphy on the following questions. Note that these are questions which do not have 'right' or 'wrong' answers. Nonetheless is should be possible for an experienced scientist or technologist to distinguish between a 'reasonable' answer and an 'unreasonable' one.

(a) What will be the average yield of future batches?

(b) What are the highest and lowest yields we are likely to get in future batches?

(c) Can we be confident that the average yield of future batches will lie between 650
and 750 kg?

Note: worked solutions for the exercises are to be found in section 1.10. It is
advisable to read the solution for Exercise 1.1 before proceeding.

If we are to use the data in Table 1.1 to assess the average performance of the
process we would be wise to take account of the variability in the data. This
variability in the yield and impurity results from variation in the raw materials and
variation in the production process. It might be possible to reduce the batch-to-batch
variation by obtaining raw materials of more consistent quality and/or by improving
the process. However, such changes are not possible in the short term. We shall,
therefore, accept the plant and the materials as they are and try to cope with the
inherent variability.

Obviously, when analysing data, we need some way of measuring or quantifying
its variability. This can be achieved by calculating the *range* of the data:

range = largest value − smallest value

The largest yield in Table 1.1 is 741 kg and the smallest is 643 kg. Thus the range is
98 kg. Few people have difficulty understanding the range of a set of data. However,
the meaning of this simple concept is further clarified if we draw a dot plot. The range
of the data is the width of the dot plot (Fig. 1.1).

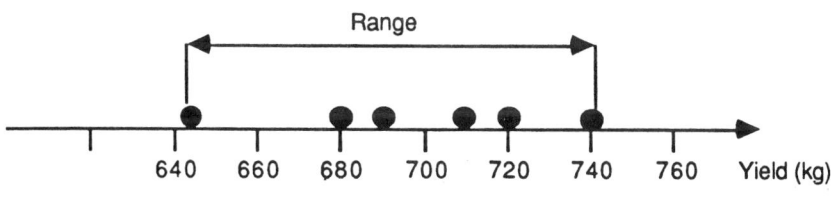

Fig. 1.1 —A dot plot.

Despite the obvious attraction of its simplicity, the range is not often used as a
measure of variability. Now that we have powerful but inexpensive calculators we
prefer to use an alternative known as the **standard deviation**. Unfortunately the
standard deviation is not as easy to understand as the range, but it is superior in
several respects. For example, use of the standard deviation overcomes two
difficulties which accompany the use of the range:

(a) A larger sample is very likely to have a greater range than a smaller sample.
(b) The range of a set of data is very much influenced by the smallest and largest

values and may not truly reflect the variability in the data. This point is illustrated in Exercise 1.2.

Before we analyse the data in Table 1.1, I would like you to assess the variability of several sets of data. To carry out this exercise you will require a calculator which has a built-in routine for calculating standard deviations. While using such a calculator you do not need to know how a standard deviation is calculated. However, if you are interested in the details of the calculation they are given in Appendix A. Your calculator probably has two keys which give standard deviations. It is the larger of the two results that you require. Appendix A explains the difference. The handbook for your calculator probably uses the word 'mean' rather than 'average'. To avoid confusion I shall do the same.

Exercise 1.2
(a) Calculate the mean, the range and the standard deviation for each of the following sets of data. They give the weights in kg of six items.

     (i) 4, 1, 3, 4, 3, 6
    (ii) 4, 1, 3, 2, 5, 6
   (iii) 1, 6, 1, 1, 6, 6
   (iv) 1, 6, 1, 1, 1, 1
    (v) 6, 6, 1, 6, 6, 6

(b) Draw a rough dot plot for each set of data in part (a). Write next to each dot plot the mean, the range and the standard deviation.
(c) Compare your dot plots for data sets (i), (ii) and (iii). Do you agree that the standard deviation gives a better indication of variability than does the range?

The standard deviation of the yields in Table 1.1 is 34.502 kg. The mean yield is 698 kg. We shall now make use of these two numbers to estimate the mean yield of future batches and to indicate the precision of our estimate. There are many ways in which this can be done, but the most useful perhaps is to calculate what are known as **confidence limits**.

95% confidence limits for the mean yield of future batches are given by:

$$\bar{x} \pm ts/\sqrt{n}$$

where $\bar{x}$ is the mean yield of the six batches, s is the standard deviation of the six yields, n is the number of batches (i.e. 6) and t is from the 95% column of Table ST1 with $(n-1)$ degrees of freedom.

Table ST1 is the first of the *statistical tables* which are located near the end of the book. As n is equal to 6 we look for 5 degrees of freedom in the left-hand column of Table ST1. We see that the value of t is 2.57.

Using $\bar{x} = 698$, $s = 34.502$, $n = 6$ and $t = 2.57$ we obtain

$$698 \pm 2.57 \, (34.502)/\sqrt{6}$$
$$= 698 \pm 36.2$$
$$= 661.8 \text{ to } 724.2 \, \text{kg}$$

The lower confidence limit is 661.8 kg and the upper confidence limit is 734.2 kg. We can be 95% confident that the mean yield of future batches will lie between these two limits. The range of values between 661.8 and 734.2 kg is known as a **confidence interval**.

You may be suprised that the confidence interval is so wide. However, the width of this interval is a true reflection of the variability from batch to batch and of the limited amount of data on which the confidence limits are based. If Dr Murphy wished to estimate the mean yield of future batches with greater precision he would need to acquire more data by producing additional batches. The confidence interval tends to become narrower as the number of batches increases. This point is illustrated in Exercise 1.3.

Exercise 1.3

Suppose Dr Murphy was so anxious to assess the effectiveness of his modification that he attempted to estimate the mean yield of future production after only two **batches had been produced**.

(a) Calculate the mean and standard deviation of the first two yields in Table 1.1.
(b) Use the mean and standard deviation from part (a) to calculate 95% confidence limits for the mean yield of future batches.
(c) Why is your confidence interval wider than the interval we obtained earlier from the data on all six batches?

Repeating the calculation of confidence limits using three, four and then five batches gives us the confidence intervals in Table 1.2 which are displayed in Fig. 1.2. It is clear that the width of the interval decreases as the number of batches increases. Perhaps it is also clear that the benefit of including the third batch is much greater than the benefit of adding the sixth. Because of the square root in the formula we must quadruple the number of batches in order to halve the width of the interval.

I hope that you have accepted confidence limits as a useful, and an entirely reasonable, way of indicating the precision of an estimate. However, I do realize that you may have misgivings about using a formula without some knowledge of either its origins or the assumptions on which it is based. We shall discuss these assumptions after we have explored a second situation in which confidence limits are applicable.

Table 1.2 — Confidence intervals for mean yield of future batches

Number of batches produced	Mean yield (kg)	Standard deviation (kg)	Degrees of freedom	Value of t from Table ST1	Confidence interval for mean yield of future batches
2	705.5	21.92	1	12.71	705 ± 197
3	684.7	39.27	2	4.30	685 ± 97
4	698.8	42.68	3	3.18	699 ± 68
5	695.2	37.80	4	2.78	695 ± 47
6	698.0	34.50	5	2.57	698 ± 36

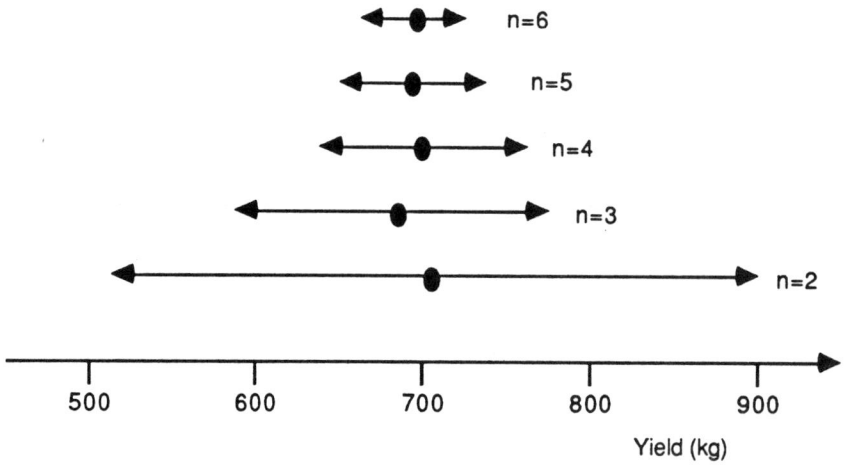

Fig. 1.2 — Confidence intervals from Table 1.2.

1.3 ASSESSING THE QUALITY OF A CONSIGNMENT

John Rogers is the Quality Manager of Cocalite Limited, a small company which manufactures various chemical intermediates using coal as the basic feedstock. Among John's many responsibilities is the assessment of the quality of all raw materials purchased by Cocalite. He is currently very concerned with the determination of sulphur content of coal supplies. This has proved troublesome in the past but a new rapid analytical method has recently become available. A 100 tonne consignment of coal is assessed by taking ten lumps and measuring the sulphur content of each using the new method. The results are given in Table 1.3.

The mean of the ten determinations is 0.341% and the standard deviation is

Table 1.3 — Ten determinations of sulphur content

0.36%	0.28%	0.33%	0.33%	0.39%
0.30%	0.41%	0.35%	0.30%	0.36%

0.041 22%. We can use these results to calculate confidence limits for the true sulphur content of the consignment. What exactly do we mean by the expression 'true sulphur content'? It is the mean sulphur content that we would have obtained if we had examined every lump of coal in the consignment. Using $\bar{x} = 0.341$, $s = 0.4122$, $n = 10$ and $t = 2.26$ we obtain the following 95% confidence limits for the mean sulphur content of the consignment:

$$\bar{x} \pm ts/\sqrt{n}$$
$$= 0.341 \pm 2.26\,(0.041\,22)/\sqrt{10}$$
$$= 0.341 \pm 0.029$$
$$= 0.312\% \text{ to } 0.370\%$$

Thus we can be 95% confident that the true sulphur content of the coal consignment lies between 0.312% and 0.370%.

You will realize, of course, that these confidence limits might have been different. Suppose, for example, that the chemist who selected the ten lumps of coal from the consignment had entrusted this task to his laboratory assistant. Almost certainly ten different lumps of coal would have been selected, giving ten different determinations, a different mean and a different standard deviation. If the mean had been 0.345% and the standard deviation 0.017 29%, for example, the confidence limits would have been 0.333% and 0.357%. This interval is narrower because the standard deviation is smaller. If several people each selected ten lumps of coal from the consignment the resulting confidence intervals might give a diagram similar to Fig. 1.3.

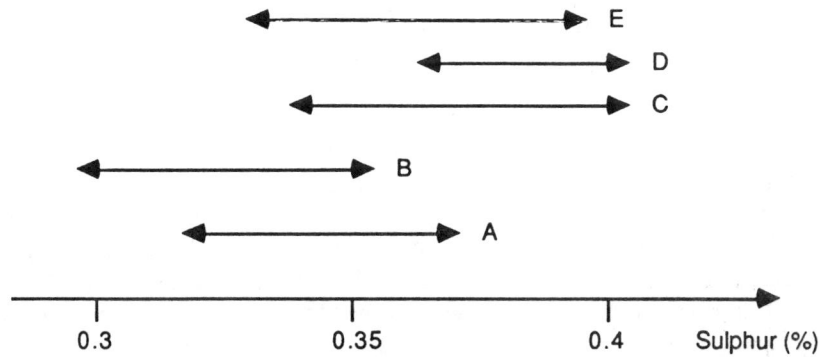

Fig. 1.3 — Several estimates of the true sulphur content.

Each person who produced a confidence interval in Fig. 1.3 felt confident that the true sulphur content of the consignment was within this interval. If the true sulphur content ever became known, would any of the six be disappointed? Without knowing the true sulphur content we can answer 'yes', because intervals B and D do not overlap. Thus it is impossible for both B and D to be correct. This statement is not intended as a criticism of either person. Perhaps one or other obtained a sample of coal which was not representative of the whole consignment. You might wonder how anyone would end up with an unrepresentative sample. It might result from a bad sampling method or from bad luck. A regular user of 95% confidence intervals can expect bad luck to strike on one occasion in twenty. With every 95% confidence interval there is 95% chance of being correct, but there is a 5% chance (i.e. 1 in 20) of being wrong. If you wished to reduce this chance of error you could calculate 99% confidence limits. The same formula is used but the value of t is taken from the 99% column of Table ST1.

Exercise 1.4

(a) Using the data in Table 1.3, calculate 99% confidence limits for the mean sulphur content of the whole consignment of coal.

(b) Use the data in Table 1.3 to calculate 99.9% confidence limits for the mean sulphur content of the consignment of coal.

(c) If 1000 people each took a sample of coal and calculated 99% confidence limits for the sulphur content of the consignment, how many would you expect to have the true percentage of sulphur within the interval?

(d) When you calculate 99.9% confidence limits for the true sulphur content what is the chance that your confidence interval will not include the true value?

1.4   ASSESSING OTHER AVERAGES

We have just explored some of the difficulties associated with assessing the sulphur content of a consignment of coal. Earlier we discussed the problem of assessing the yield of a production process. In both cases it was an **average** value that we wished to assess, and in both cases we calculated confidence limits between which we hoped the average would lie.

Obviously this technique could be used in many other situations. For example, a pig farmer wishes to assess the average weight gain of pigs fed on an experimental diet. Clearly he can feed the diet to only a small number of pigs. However, he can then use the data to estimate the mean weight gain that he would have found if he had fed the diet to **all** pigs of that breed. As a further example, consider the Company Medical Officer who wishes to assess the average blood pressure of all personnel who have worked on a particular process (perhaps it is now suspected that the process emitted dangerous radiation in the past). Initially he measures the blood pressure of a small number of workers and he uses these data to estimate the blood pressure of **all** workers who have been associated with the process.

These two examples, and the two considered earlier, have certain features in common. Let us now focus on these essential features so that you will easily recognize other situations in which confidence limits might be appropriate

In all four situations the researcher has taken a **sample** from a **population**. He has then used the sample mean to estimate the population mean.

As an industrial scientist or technologist you probably use the word 'sample' quite often, but you may be less happy with 'population'. Statisticians, however, use both words, frequently and inseparably. In fact you could define statistics as a body of knowledge which can be useful to anyone who has taken a sample from a population. The two words are so important in statistics and data analysis that it would be wise to define them without further ado.

A **population** is a large number of items (or a large quantity), while a **sample** is simply a smaller number of items (or smaller quantity). The sample is taken from the population so that the data from the sample can be used to draw conclusions about the population. Obviously there is a risk that the conclusions may be incorrect or misleading, but we tolerate this risk because it is less costly to examine the sample than to examine the whole population.

'Why', you may ask, 'does the statistician have this obsession with populations?' The answer is simple. Many statistical techniques are valid only if:

(a) the population has certain characteristics, and
(b) the sample is selected from the population in a particular way.

Thus it is not possible to use statistical techniques with safety unless you pay some attention to the population as well as the sample. With this sobering thought in mind let us re-examine some of the situations described earlier to see whether we can identify the sample and the population in each.

Dr Murphy measured the yield of six batches of pigment in order to estimate the mean yield of future batches. His sample consists of the first six batches produced after the modification and his population consists of all batches made after the modification plus all future batches. As in all statistical analysis, the data come from the sample but the conclusions refer to the whole population.

John Rogers gathered the data in Table 1.3 in order to assess the sulphur content of a 100 tonne consignment of coal. His sample consists of ten lumps of coal. His population is much larger; it is the whole consignment. He used the mean sulphur content of the sample to estimate the mean sulphur content of the population. The advantage of analysing only ten lumps of coal rather than 100 tonnes is obvious, but this must be weighted against the risk of getting a misleading result.

Exercise 1.5

The Secretary of the Royal Society of Chemistry wished to ascertain the views of the members of the Society on a particular issue. To reduce the cost of this investigation he decided to contact only 10% of the members in the hope that their views would be representative of the views of all members. He therefore sent a questionnaire to every tenth person on the membership list.

(a) Define the **population**, about which the Secretary wished to draw conclusions.
(b) Define the **sample**, from which the Secretary obtained his data.
(c) If every member of the society had **exactly the same** opinion on this issue, how many members would need to be included in the sample in order to ascertain this opinion precisely?

1.5 DO WE NEED SUCH A LARGE SAMPLE!

Common sense tells us that a larger sample is likely to give a better estimate of the population mean than we would obtain from a smaller sample. With a larger sample any unusually high value is likely to be counterbalanced by low values, thus giving a sample mean close to the population mean. In contrast, with a small sample just one peculiar value can have large influence on the sample mean, causing it to deviate considerably from the population mean. So it is better to have a larger sample.

Obviously the larger sample is more expensive. Furthermore, as we noted earlier, it is necessary to quadruple the sample size in order to halve the width of the confidence interval. Thus it may be very wasteful to increase the size of the sample beyond a certain point. In practice we require a sufficiently accurate estimate at a cost we can afford, and we wish to know, in advance, how large a sample is needed to give us this accuracy.

The size of sample needed to give confidence limits of $\pm c$ is given by:

$$n = (ts/c)^2$$

where n is the required sample size, c is the width of interval we hope to obtain, s is a standard deviation which quantifies the variability of the situation and t is taken from Table ST1 with appropriate degrees of freedom.

Let us use this formula to calculate how large a sample Dr Murphy would require if he wished to estimate the mean yield of future batches to within ± 10 kg. First we need a standard deviation which indicates the batch-to-batch variation in yield. The only data available are in Table 1.1. The standard deviation of the six yield values in Table 1.1 is 34.502 kg which has 5 degrees of freedom. From Table ST1, with 5 degrees of freedom and 95% confidence, we obtain a t value of 2.57. Putting c equal to 10 we can proceed with the calculation:

$$n = (ts/c)^2$$
$$= [2.57(34.502)/10]^2$$
$$= 78.6$$

Obviously the sample size must be a whole number, so we round up this result to 79.

Thus Dr Murphy would require at least 79 batches of digozo blue pigment in order to estimate the yield of future batches to within ± 10 kg. This figure of 79 is an estimate of course. It is based on the data in Table 1.1, which is the only information available. If we had more data we could calculate a better estimate of how many batches were needed. You could say that this is a 'chicken and egg' formula in one respect, since it cannot be used to estimate the required sample size unless we already have some data to give us a standard deviation. Thus, for example, if you have a new process it is impossible to say how many runs are needed to estimate its average performance, unless you already have some knowledge of its variability.

Exercise 1.6

Table 1.3 contains ten determinations of sulphur content. Each determination was based on one lump of coal taken from a 100 tonne consignment.

(a) Use the standard deviation of the ten determinations to calculate the number of lumps that would be needed to estimate the sulphur content of the consignment to within $\pm 0.01\%$.
(b) Suppose the person who selected the ten lumps from the consignment was able to judge the approximate sulphur content of a lump of coal from its appearance. Suppose also that he had deliberately chosen 10 lumps which appeared very similar to each other. Would you wish to increase or decrease the sample size calculated in part (a)?

1.6   PREDICTING THE YIELD OF INDIVIDUAL BATCHES

From the yields of the six batches in Table 1.1, Dr Murphy has concluded that the mean yield of future batches will lie between 661.8 kg and 734.2 kg. Despite the great width of this interval he decided to include the confidence limits in his report rather than simply to quote the mean of the six batches. He feels that the width of the confidence interval gives a fair indication of the precision of his estimate of the long-term average yield for the digozo process.

Unfortunately at least one reader of his report does not fully understand the idea of confidence limits for, in a discussion of Dr Murphy's report, the question is asked 'Does this imply that 95% of future batches will have a yield between 661.8 kg and 734.9 kg?' The simple answer to this question is 'no". Confidence limits for the population mean tell us very little about what we are likely to find when we examine individual batches. We are confident that the long-term mean yield lies between 661.8 kg and 734.9 kg, but the yields of individual batches will be scattered around this mean. Some batches will be well above the mean and some will be well below. If we wish to discuss individual batches we should calculate what are known as **tolerance limits**.

Tolerance limits for a specified percentage of a population are given by:

$$\bar{x} \pm k\dot{s}$$

where $\bar{x}$ is the sample mean, s is the sample standard deviation and k is taken from Table ST2.

Let us suppose that Dr Murphy wishes to include in his report a prediction that 'Almost all batches will have a yield between X kg and Y kg'. If we interpret 'almost all' to be 90% and we wish to be 95% confident that the prediction is correct then we use a value of 3.71 from Table ST2. To carry out the calculation of tolerance limits we also need the mean and standard deviation of the six batches in Table 1.1. These are 698 kg and 34.502 kg.

$$\bar{x} \pm ks = 698 \pm 3.71(34.502)$$
$$= 698 \pm 128.0$$
$$= \text{568.? TO 1434 KG} \quad 570 \text{ TO } 826 \text{ KG}$$

Thus we can be 95% confident that almost all batches of digozo blue pigment will have yields between 570 kg and 826 kg. You can see that this tolerance interval is much wider than the confidence interval for the mean yield. The difference between the two intervals is illustrated in Fig. 1.4.

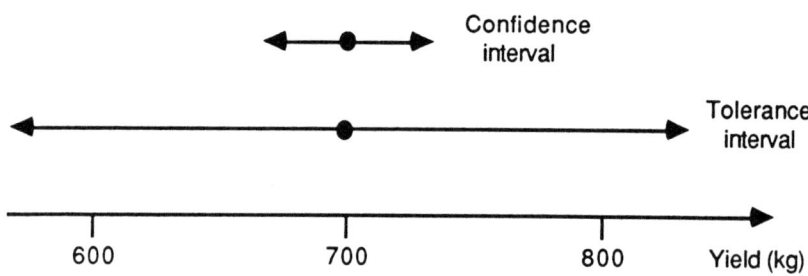

Fig. 1.4 — Confidence interval and tolerance interval.

Exercise 1.7
Each sulphur determination in Table 1.3 is based on **one** lump of coal. The purpose of sampling the ten lumps was to estimate the mean sulphur content of the whole consignment, but we might wish to speculate about the variation in sulphur from lump to lump.

(a) Calculate 95% tolerance limits for 90% of lumps in this consignment.
(b) Compare the tolerance interval with the 95% confidence interval obtained earlier, 0.312% to 0.370%.

(c) (More difficult) Both the tolerance interval and the confidence interval would be narrower if we had a larger sample. How wide would each interval be if we measured the sulphur content of **every** lump in the 100 tonne consignment?

1.7 STATISTICAL TECHNIQUES ARE DANGEROUS

While studying this chapter, you should have become aware of the benefits of using the statistical techniques presented. The use of these techniques can enable the scientist to enjoy the economies that result from sampling. Unfortunately the use of these techniques can also lead the scientist to false conclusions. In fact there are two risks involved.

(a) the risk of drawing a false conclusion because your sample is not representative of the population from which it was drawn;
(b) the risk of drawing a false conclusion because you are violating the assumptions underlying the statistical technique.

The sampling risk is unavoidable, but it can be minimized by following reputable sampling procedures. Furthermore, the sampling risk can be quantified. For example, if we calculate 95% confidence limits for a population mean we run a 5% risk that the true value will be outside the confidence interval.

The risk of drawing a false conclusion as a result of violating an assumption cannot be quantified. However, the risk must be much greater for those who are unaware of the assumptions, so let us examine them briefly. A fuller discussion of the underlying assumptions will be given in a later chapter.

While calculating confidence limits and tolerance limits we made use of two statistical tables, Tables ST1 and ST2. There are many other statistical tables which will be of use in other chapters. Each table is associated with a specific statistical technique and each table has certain assumptions built into it. Thus it is impossible to circumvent these assumptions. Tables ST1 and ST2 share two assumptions:

(a) that the sample was selected at random from the population;
(b) that the population has a normal distribution.

Statisticians use the word 'random' with great care. They define **random sampling** as 'a process which ensures that every member of the population has the same chance of being included in the sample'. The end product of random sampling is a random sample.

Unfortunately there are many situations in which we cannot obtain a random sample. For example, the Plant Manager who wished to assess the mean yield of the digozo process could not use random sampling. It was the yield of future batches that he was really interested in, but he could not sample from batches not yet produced. Clearly the sampling assumption is violated, but he may be happy to calculate confidence limits if he believes that the six batches he has produced are **representative** of future batches.

The normal distribution assumption is also violated in many situations, but not severely one hopes. A full discussion of the normal distribution must be reserved for a later chapter but a comparison of two sets of data will illustrate its major feature. Fig. 1.5 contains dot plots for the yield data and the impurity data from Table 1.1.

Fig. 1.5 — Yield and impurity of six batches.

There are so few data in Fig. 1.5 that it would be dangerous to predict what the dot plots would look like if the Plant Manager had produced a larger number of batches. It would be even more dangerous to predict the distribution of yield and impurity for the **whole population** of future batches. Nonetheless, I have done so in Fig. 1.6.

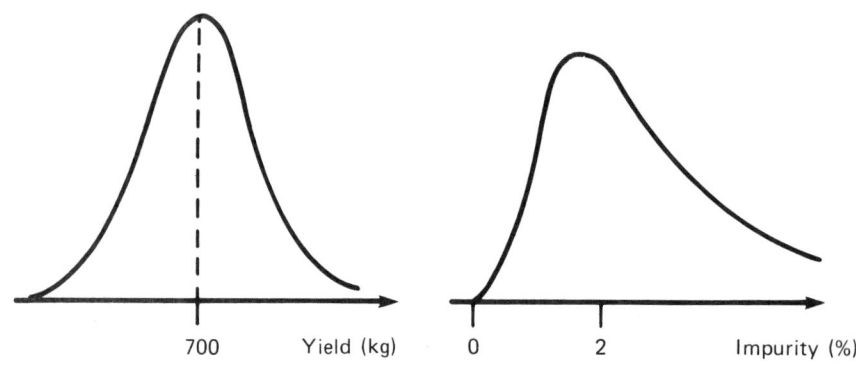

Fig. 1.6 — Yield and impurity of the population of batches.

In Fig. 1.6 I have not drawn a dot for each individual batch in the population. Because there are so many batches I have drawn a smooth curve which encompasses all the dots. The distinctive, bell-shaped curve on the left is a normal distribution curve. It is **symmetrical** about the mean. The very different curve on the right is obviously not symmetrical, but **skewed**. I am suggested that the impurity data did **not** come from a normal distribution. Thus it would **not** be wise to calculate confidence limits for the mean impurity.

You may be alarmed by the way I have introduced the two curves in Fig. 1.6, supported only by the flimsy evidence in Fig. 1.5. However, there are well-proven methods for answering the question 'Did this set of data come from a normal distribution?' We shall discuss these methods in a later chapter.

1.8   A SUMMARY OF THE IMPORTANT POINTS

(1) In this chapter we have used data form a sample to **estimate** certain features of the population from which the sample was taken. We have focused particularly on the problems associated with estimating the population mean.

(2) A **population** is simply a large number of items or a large quantity. The **sample** is a smaller quantity. We take a sample because it would be uneconomical to examine the whole population.

(3) Whenever we take a sample we run the **risk** that our conclusions may be misleading because the sample is not representative of the population from which it was drawn. One assumption underlying many statistical techniques is that **random sampling** was used.

(4) When we examine the data from our sample we will almost certainly find **variability** from item to item. The variability within the sample is simply a reflection of variability within the population. Because of this variability it is most unlikely that any sample will be perfectly representative of the population.

(5) When estimating the population mean it is wise to take account of the variability in the data. This can be achieved by calculating the sample standard deviation and using this to obtain **confidence limits** for the population mean using the formula:

$$\bar{x} \pm ts/\sqrt{n}$$

where $\bar{x}$ is the sample mean, s is the sample standard deviation, n is the sample size and t is from Table ST1 with appropriate degrees of freedom.

(6) A **larger sample** will tend to give a narrower confidence interval than we would obtain from a smaller sample. Thus, increasing the sample size is likely to bring us closer to the truth, provided that our sampling method is not biased.

(7) The size of sample required to estimate the population mean to within $\pm c$ can be estimated from:

$$n = (ts/c)^2$$

where s is a suitable standard deviation and t is taken from Table ST1, with appropriate degrees of freedom.

(8) Note the distinction between a confidence interval and a tolerance interval. The confidence interval tells us a lot about the population mean, but it tells us nothing about individual observations.

1.9   ADDITIONAL EXERCISES

Exercise 1.8

Piscean Products Limited own several trout farms in the north of England. The

Research Director, Dr Gill, has initiated several research studies to compare the effectiveness of different diets and feedings schedules. The fish in one particular tank have been fed on diet TX7 since their introduction six weeks ago. 16 fish taken from this tank are found to have the following weights, in grams:

```
548  641  597  624  651  582  590  614
610  565  580  603  620  607  611  597
```

(a) Calculate the mean and standard deviation of the 16 weights.
(b) Calculate 95% confidence limits for the mean weight of **all** the trout in the tank from which these 16 were taken.
(c) Use the standard deviation from part (a) to estimate the number of trout that would need to be weighed in order to estimate the population mean weight to within $\pm 5\,g$.
(d) Calculate 95% tolerance limits for the weights of 90% of trout fed on this diet.
(e) Explain to Dr Gill the difference between the confidence limits for part (b) and the tolerance limits from part (d).
(f) Draw a dot plot of the weights of the 16 fish.
(g) Does the dot plot indicate that the assumptions underlying Tables ST1 and ST2 are violated?

Exercise 1.9

Statisticians have their own peculiar language, as do chemists, engineers, biologists etc. Some acquaintance with statistical terminology is essential if you wish to report your findings clearly or to understand the reports of other scientists. It is certainly not my intention to inflict on you more than the minimum of statistical jargon, but I would like you to understand the terminology introduced in this chapter before you proceeed to Chapter 2. To test your understanding of this terminology, insert the missing words in the following passage.

We define **statistics** to be 'a body of knowledge which can be of use to anyone who has taken a (1) _____ from a (2) _____ in which there is (3) _____ from item to item'. With a hypothetical population in which there is no variability, the population standard deviation would be equal to (4) _____ and every measurement on an item in the sample would be equal to the population (5) _____. With such a population the researcher would need only one member of the population in his sample. In practice, of course, the researcher finds variability from item to item within his sample. This is simply a reflection of variability within the (6) _____ from which the sample was taken. The presence of (7) _____ within the sample implies that the sample mean will not give us a perfect estimate of the (8) _____ mean. For this reason the scientist may prefer to quote two numbers rather than a single estimate for the population mean. These two numbers are known as confidence (9)

and the range of values between the two numbers is known as
a (10) _____ (11) _____ . The width of a confidence interval
depends on two factors:

(a) the (12) _____ from item to item within the sample.
(b) the (13) _____ of items in the sample.

A larger sample can be expected to give a (14) _____ confidence
interval. In contrast, we would expect a wider confidence interval if the (15)
_____ within the sample were greater.

1.10   WORKED SOLUTIONS

Solution to Exercise 1.1

There are no correct answers to this exercise. If your solutions differ from mine, do
not assume that you are wrong, or that I am wrong. Reconsider your answer in the
light of my comments.

(a) No one can predict with certainty the average yield of future batches. Nonethe-
    less, it is possible to make a reasonable estimate of the long-term average yield
    from the information in Table 1.1. The average yield of the six batches is 698 kg.
    It is reasonable to suggest that future batches will give an average yield close to
    698 kg — but how close? We must bear in mind that

    (i) there is variation in yield from batch to batch within Table 1.1, and
    (ii) there are very few data in Table 1.1.

    If we had data from a larger number of batches and if there was less variability
    from batch to batch, we could have more confidence in our predictions.
(b) We see that among the six batches, the lowest yield was 643 kg and the highest
    yield was 741 kg. However, it is quite possible that, in the future, we will obtain
    batches which yield less than 643 kg and batches which yield more than 741 kg.
    Thus it would be foolish to predict that all future batches would have yields
    within the range 643 to 741 kg. We need to quote a wider range — but how wide?
    I shall return to this problem later in the chapter.
(c) For forward planning, the Plant Manager is not particularly interested in
    individual batches. He is more concerned with the long-term average yield. Can
    he be confident that the average yield of future batches will lie between 650 kg
    and 750 kg? There is a difference of 100 kg between these two figures, but can we
    be certain that the mean yield will lie in this range? If we quoted two numbers
    which were further apart we could be more confident. This problem will form the
    central theme of this chapter.

Solution to Exercise 1.2

(a)   Mean   Range   Standard deviation

   (i)   3.5 kg   5.0 kg   1.643 kg
  (ii)   3.5 kg   5.0 kg   1.871 kg
 (iii)   3.5 kg   5.0 kg   2.739 kg
 (iv)   1.83 kg   5.0 kg   2.041 kg
  (v)   5.17 kg   5.0 kg   2.041 kg

(b)   (i)   Mean   = 3.5 kg
           Range   = 5.0 kg
           Standard deviation   = 1.643 kg

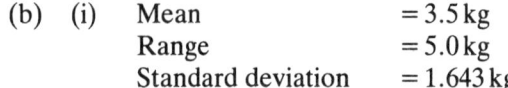

Fig. 1.7 — Dot plot for the data in (i).

  (ii)   Mean   = 3.5 kg
           Range   = 5.0 kg
           Standard deviation   = 1.871 kg

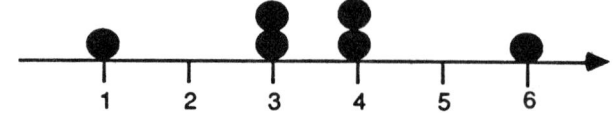

Fig. 1.8 — Dot plot for the data in (ii).

 (iii)   Mean   = 3.5 kg
           Range   = 5.0 kg
           Standard deviation   = 2.739 kg

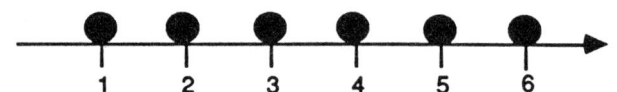

Fig. 1.9 — Dot plot for the data in (iii).

 (iv)   Mean   = 1.83 kg
           Range   = 5.0 kg
           Standard deviation   = 2.041 kg

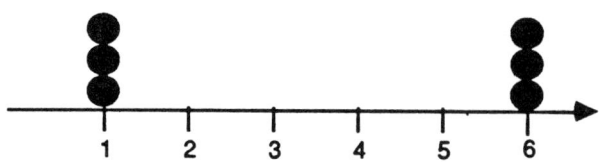

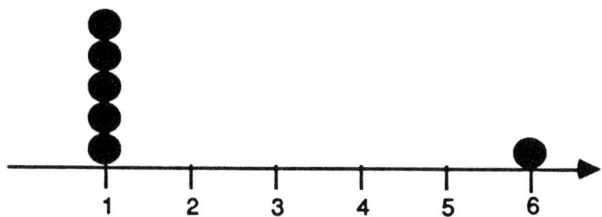

Fig. 1.10 — Dot plot for the data in (iv).

(v) Mean   = 5.17 kg
     Range   = 5.0 kg
     Standard deviation   = 2.041 kg

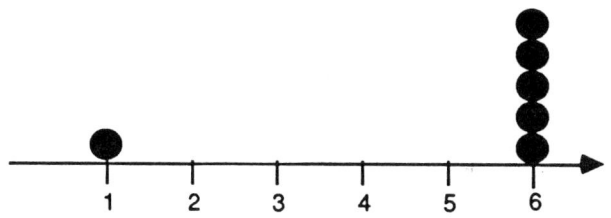

Fig. 1.11 — Dot plot for the data in (v).

(c) The dot plots indicate that the data in (iii) are more widely spread or scattered than the data in (i). This greater variability is reflected in the standard deviation, but not in the range. The standard deviation gives a better indication of spread than does the range.

Solution to Exercise 1.3
(a) For the first two batches in Table 1.1 the yield is 705.5 kg and the standard deviation is 21.92 kg.
(b) To calculate 95% confidence limits for the long-term mean yield we need a value of t from Table ST1. The standard deviation of two observations has only 1 degree of freedom, so the value of t will be 12.71.

$$\bar{x} \pm ts/\sqrt{n}$$
$$= 705.5 \pm 12.71\,(21.92)/\sqrt{2}$$
$$= 508.5 \text{ to } 902.5 \text{ kg}$$

(c) This confidence interval is wider than the interval we calculated earlier using data from six batches. We can see in the formula that three factors contribute to the width of a confidence interval.

  (i) With only 1 degree of freedom we have a very large value for t. This gives a wider interval.

(ii) A smaller sample gives a smaller value of n and thus a wider interval.

(iii) A larger standard deviation will give a wider interval. However, the first two batches have a smaller standard deviation (21.92) than the first six batches (34.50), so this is not the explanation for the wider interval calculated in part (a).

Solution to Exercise 1.4

(a) For the ten sulphur determinations in Table 1.3 the mean is 0.34% and the standard deviation is 0.4122%. With 9 degrees of freedom the t value for 99% confidence is 3.25:

$$\bar{x} \pm ts/\sqrt{n}$$
$$= 0.341 \pm 3.25(0.04122)/\sqrt{10}$$
$$= 0.341 \pm 0.042$$
$$= 0.299\% \text{ to } 0.383\%$$

(b) With 9 degrees of freedom the t value for 99.9% confidence is 4.78:

$$\bar{x} \pm ts/\sqrt{n}$$
$$= 0.341 \pm 4.78 \, (0.04122)/\sqrt{10}$$
$$= 0.279\% \text{ to } 0.403\%$$

(c) We could expect 99% of the people to produce confidence intervals which included the true percentage of sulphur. Thus we would expect 990 people to be correct and only 10 people to produce intervals which did not include the true value.

(d) When you calculate 99.9% confidence limits, there is 99.9% chance that your confidence interval will contain the true sulphur content. Thus there is only a 0.1% chance, or 1 in 1000 chance, that your interval will not contain the true value. Obviously we can reduce the risk of being wrong by increasing the level of confidence, which widens the interval.

Solution to Exercise 1.5

(a) The population consists of **all** members of the Royal Society of Chemistry.

(b) The sample consists of the 10% of members who are selected from the list.

(c) If all members of the society held exactly the same opinion he would need to contact **only one** member. If there is no variability within the population each member of the population can be regarded as a representative sample. In practice, of course, the scientist may be confronted with considerable variability and a large sample enables him to 'average out' the variation.

Solution to Exercise 1.6

(a) $c = 0.01$
        $s = 0.04122$
        $t = 2.26$
        $n = (ts/c)^2$
        $= [2.26(0.04122)/0.01]^2$
        $= 86.8$

Thus we would need a sample of at least 87 lumps in order to estimate the sulphur content of the consignment to within $\pm 0.01\%$.

(b) The person who selected the ten lumps of coal used his judgment, rather than selecting 'at random'. By so doing he may have introduced a bias, thus increasing the chance that his sample mean is very different from the population mean. Regardless of bias, he will certainly have **reduced the variability** in the sample by 'deliberately choosing 10 lumps which appeared very similar to each other'. Thus his sample standard deviation will probably be less than the population standard deviation, and the calculation carried out in part (a) will have **underestimated** the size of sample required by someone using conventional sampling methods.

Solution to Exercise 1.7

(a) Using a confidence level of 95% and a sample size of 10 the value of k for 90% of the items is 2.84, from Table ST2. The mean and standard deviation of the data in Table 1.3 are 0.341% and 0.3122%.

$$\bar{x} \pm ks = 0.341 \pm 2.84(0.041\,22)$$
$$= 0.341 \pm 0.117$$
$$= ?.224\% \text{ to } 0.458\%$$

(b) The tolerance interval calculated in part (a) is wider than the confidence interval. A tolerance interval for a specified percentage of individual items will always be wider than a confidence interval for the population mean.

(c) If we measured the sulphur content of every lump of coal in the consignment the confidence interval would have zero width. The tolerance interval would not have zero width. Its width would depend on the size of the population standard deviation.

(Note that the formula we have used for confidence limits is only appropriate if the population is much larger than the sample. When the sample is more than 10% of the population we should modify the formula to

$$\bar{x} \pm ts\sqrt{[(1/n) - (1/N)]}$$

where N is the size of the population.)

Solution to Exercise 1.8

(a) Mean weight = 602.5 g; standard deviation of weights = 26.336 29 g.

(b) The standard deviation calculated in part (a) has 15 degrees of freedom. From Table ST1, with 15 degrees of freedom and 95% confidence, the value of t is 2.13.

$$\bar{x} \pm ts/\sqrt{n} = 602.5 \pm 2.13\,(26.336\,29)/\sqrt{16}$$
$$= 602.5 \pm 14.0$$
$$= 588.5\,g \text{ to } 616.5\,g$$

Thus we can be 95% confident that the mean weight of all fish in this tank will lie between 588.5 g and 616.5 g.

(c) $c = 5$

$t = 2.13$

$s = 26.33629\,\text{g}$

$n = (ts/c)^2$

$\quad = [2.13(26.33629)/5]^2$

$\quad = 125.9$

Thus we would need to weigh at least 126 trout from this tank in order to estimate the population mean weight to within $\pm 5\,\text{g}$.

(d)　　　$\bar{x} = 602.5\,\text{g}$

$s = 26.33629\,\text{g}$

$k = 2.44$

$\bar{x} \pm ks = 602.5 \pm 2.44\,(26.33629)$

$\qquad\quad = 602.5 \pm 64.3$

$\qquad\quad = 538.2 \text{ to } 666.8\,\text{g}$

Thus we can be 95% confident that 90% of fish within this tank will have weights between 538.2 g and 666.8 g.

(e) The confidence limits tell us that the mean weight of the fish in this tank almost certainly lies between 588.5 g and 616.5 g. Whatever the mean actually is, we can be sure that many fish will have weights greater than the mean and many will have weights less than the mean. The tolerance limits tell us that 90% of fish in this tank will have weights between 538.2 g and 666.8 g. A tolerance interval tells us something about the weights of **individual** fish, whereas a confidence interval tells us something about the **mean**, but nothing about individual fish.

(f)

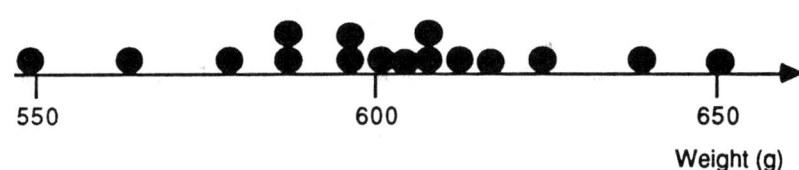

Fig. 1.12 — Weights of 16 fish on diet TX7.

(g) The dot plot is roughly symmetrical, indicating that the weights of fish in this tank could well have a normal distribution. The dot plot tells us **nothing** about how the 16 fish were selected from the thousands of fish in the tank. Thus the dot plot does not help us to check the random sampling assumption. Common sense suggests that our conclusions could be misleading if, for example, the technician selected all 16 fish from the same part of the tank or if he selected those fish which were easiest to catch.

Solution to Exercise 1.9

(1) sample

(2) population

(3) variability or variation

(4) zero

(5) mean

(9) limits

(10) confidence

(11) interval

(12) variation or variability

(13) variation or variability

(6) population   (14) narrower
(7) variation or variability   (15) variability
(8) population

1.11   DETAILED OBJECTIVES FOR THIS CHAPTER

Now you have studied the material in this chapter and attempted to relate its content to your existing knowledge, you should be able to do the following.

(1) Explain the meaning of the following terms and use them correctly in appropriate contexts:
   (a) sample;
   (b) population;
   (c) mean or average;
   (d) standard deviation;
   (e) variable;
   (f) confidence limits;
   (g) confidence interval;
   (h) tolerance limits;
   (i) tolerance interval;
   (j) degrees of freedom (see Appendix A).

(2) Use your pocket calculator to obtain the mean and standard deviation of a set of data.
(3) Calculate confidence limits for a population mean, making appropriate use of Table ST1.
(4) Calculate tolerance limits, making appropriate use of Table ST2.
(5) Explain how the width of a confidence interval and the width of a tolerance interval depend on the sample size and the variability within the sample.
(6) Use an appropriate standard deviation to calculate an estimate of the size of sample needed to estimate a population mean to within $\pm c$.
(7) Briefly explain the assumptions underlying the formulae and the statistical tables which are used in the calculation of confidence limits and tolerance limits.

1.12   SELF-TEST

Dr Wilshaw of Agrow Limited has developed a revolutionary foliar feed for tomato plants. He is confident that the use of this feed will increase the yield of tomatoes from plants of one particular variety. In a preliminary trial the yields in kg of 12 plants given the foliar feed were:

   3.2   4.1   4.9   4.6   3.8   4.9   4.3   3.9   4.7   5.2   3.5   4.4

These data are referred to in questions (1)–(5) below. You may wish to check your answers to earlier questions before attempting later questions. Please feel free to consult the learning material while answering the questions.

(1) Which of the following is the standard deviation of the 12 yields, to three decimal places?

(a) 0.341 kg
(b) 0.372 kg
(c) 0.584 kg
(d) 0.610 kg

(2) Which of the following is the range of the 12 yields?

(a) 1.4 kg
(b) 1.7 kg
(c) 2.0 kg
(d) 4.2 kg

(3) Which of the following are 95% confidence limits for the mean yield of all tomato plants of this variety when fed with this foliar feed?

(a) 0.36 to 8.94 kg
(b) 3.82 to 4.76 kg
(c) 3.90 to 4.68 kg
(d) 4.18 to 4.40 kg

(4) Which of the following are 95% tolerance limits for the yields of 90% of tomato plants of this variety when fed with this foliar feed?

(a) 1.63 to 6.95 kg
(b) 2.55 to 6.03 kg
(c) 2.67 to 5.91 kg
(d) 3.82 to 4.76 kg

(5) In the light of the variation in yield from plant to plant, how many plants would be needed to estimate the true mean yield to within ± 0.1 kg?

(a) 14
(b) 18
(c) 180
(d) with the given information it is not possible to estimate the required sample size, even approximately.

(6) Each of the following refers to a tolerance interval and a confidence interval for the mean which have been calculated from the same set of data. Which statement is true?

(a) The 95% tolerance interval will always be wider than the 95% confidence interval.
(b) The 95% tolerance interval will always be narrower than the 95% confidence interval.
(c) If the sample were very large the tolerance interval would be narrower than the confidence interval.
(d) If the sample were very small the tolerance interval would be narrower than the confidence interval.

(7) Three laboratory assistants are asked to estimate the mean sugar content of a

consignment of beet. Each selects 10 roots and makes a determination of sugar content on each root. The 95% confidence intervals reported by the three assistants are:

    Smith, 3.6 to 4.4;   Jackson, 3.7 to 4.7;   Brown, 4.3 to 4.7

Which of the following statements is true?

(a) The standard deviation of Brown's data is smaller than the standard deviation of Smith's or Jackson's data.
(b) It is not possible for all three confidence intervals to contain the population mean.
(c) The widths of the three confidence intervals indicate that Jackson is inferior to the other two assistants.
(d) The widths of the three confidence intervals tell us that only Brown's sample mean is equal to the population mean.

(8) Refer to the three confidence intervals in question (7). If each of the three assistants had also calculated a 95% tolerance interval for 90% of the population, which assistant would have the widest interval?

(a) It is not possible to say which tolerance interval would be the widest.
(b) Smith
(c) Jackson
(d) Brown

(9) In question (7) a population mean was referred to. What is meant by 'population mean' in this situation?

(a) the mean sugar content of the 30 beetroots examined by the three laboratory assistants;
(b) the mean of the three sample means calculated by the three assistants;
(c) the mean of the six confidence limits calculated by the three assistants;
(d) the mean sugar content of all beetroots in the whole consignment.

(10) Which of the following statements is true?

(a) If a researcher took a sample and calculated 95% confidence limits for the population mean we would expect the interval to include 95% of the population.
(b) If a researcher took a sample and calculated 95% tolerance limits for 90% of the population the interval is certain to include 95% of the population.
(c) If many researchers each took a sample and calculated 95% confidence limits for the population mean we would expect 5% of the confidence intervals to be misleading.
(d) If two researchers each took a sample and calculated 95% confidence limits for the population mean, the researcher with the larger sample would be certain to have the narrower interval.

1.13   ANSWERS TO SELF-TEST QUESTIONS

  (1) (d)
  (2) (c)
  (3) (c)
  (4) (c)
  (5) (c)
  (6) (a)
  (7) (a)
  (8) (c)
  (9) (d)
(10) (c)

2

Comparing several treatments

2.1 INTRODUCTION

Many of the experiments carried out by scientists and technologists are intended to help them make **comparisons**. For example, an agronomist does a field trial in order to compare four varieties of wheat. A medical researcher carries out a clinical trial so as to compare two drugs. A psychologist does an experiment in which he compares the pulse rates of three groups of people who are subjected to three types of stress.

In all three examples the researcher wishes to compare two or more sets of data. Each set of data was produced under different conditions, with these conditions having been deliberately changed by the experimenter. It is convenient to refer to these conditions as **treatments**. Thus in the agronomist's experiment each variety of wheat is a treatment. In the medical experiment each drug is a treatment. In the psychologist's experiment each type of stress is a treatment.

In all these comparison experiments the researcher will calculate the **mean** result for each treatment. His conclusions will be based on a comparison of these means, but he will need to take account of the variability, of course.

(1) The primary aim of this chapter is to extend the repertoire of techniques covered in Chapter 1. Whereas Chapter 1 was focused on simpler techniques, this chapter will concentrate on the more useful methods which are needed when we compare two or more products, methods or treatments.

(2) I am confident that your study of this chapter will further increase your ability to cope with the inherent variability of materials and processes when assessing performance or making comparisons.

(3) I believe that you will develop an appreciation of how the use of statistical methods can help to control the risks which must exist whenever decision are based on incomplete information.

2.2 COMPARING SIX TREATMENTS

You learned in Chapter 1 that some statistical techniques are easy to use, provided that you are equipped with a good calculator, of course. Before I introduce additional techniques I must point out that confidence intervals and tolerance intervals were included in the first chapter of this book because of their simplicity as well as their usefulness. The formulae were simple, the statistical tables were quite easy to use and the situations which gave us the data had been stripped of complexity.

Unfortunately, this happy condition cannot continue indefinitely. In this chapter and in later chapters we must grapple with more detail and we must use techniques and tables which are not so simple. However, the trend will be gradual and I hope that your confidence will grow more quickly than the difficulties. With these thoughts in mind let us examine a new problem which is similar in many ways to the problems in Chapter 1, but which is amenable to a more subtle analysis.

Oil extracted from the North Sea is invariably accompanied by water. The oil and the water are separated immediately, and then the water is returned to the sea. Unfortunately the separation process is not perfect, so the returning water is accompanied by a small quantity of oil, which contaminates the sea in the vicinity of the production platform. All oil companies are bedevilled by this problem and all carry out research into ways of reducing the pollution.

A biochemist employed by a well-known oil refining company has been asked to investigate the biodegradation of hydrocarbons in water. He is particularly interested in comparing the performance of several micro-organisms, which are known to degrade the oil and thus assist its dispersion. He carries out an experiment in which six vessels of sea water are deliberately contaminated with 25 parts per million (ppm) of hydrocarbon. Each vessel contains a different micro-organism. 14 days later the hydrocarbon levels are found to be as shown in Table 2.1.

Table 2.1 — Hydrocarbon levels after 14 days

Vessel	A	B	C	D	E	F
Determination of hydrocarbon level (ppm)	12.2 14.3	8.6 10.4 9.9	17.4 16.1	13.0 16.2	14.6 11.3 13.1	12.7 12.9
Mean	13.25	9.63	16.75	14.60	13.00	12.80
Standard deviation	1.485	0.929	0.919	2.263	1.652	0.141

What conclusions can the biochemist draw from the data in Table 2.1 concerning the ability of the six micro-organisms to reduce the level of hydrocarbon in sea water? The most effective micro-organism is the one which gives the greatest reduction in hydrocarbon. All six vessels contained 25 ppm at the start of the experiment. 14 days later all six vessels contained less than 25 ppm, with the largest decrease having occurred in vessel B. Thus there is some evidence that the organism in vessel B is more effective than the other five. However, before we draw any conclusions about the relative effectiveness of the six organisms, I would like you to give some thought to the variability in the data.

Exercise 2.1
A good way to check how well you understand a set of data, or the situation from which it came, is to select two numbers within the data and try to explain why they differ.
(a) Consider the largest and smallest determinations in Table 2.1. They are 8.6 in vessel B and 17.4 in vessel C. Write down several reasons why these two determinations differ.
(b) Consider now the two determinations in vessel A, 12.2 and 14.3. Why do they differ?
(c) What would you conclude if all 14 determinations in Table 2.1 were equal to, say, 12.50?

We noted in Chapter 1 that it is unwise to make use of the mean of a set of data without considering the variability within the data. It is equally obvious that we should not attach importance to the difference between two means without taking account of the variability within the two sets of data.

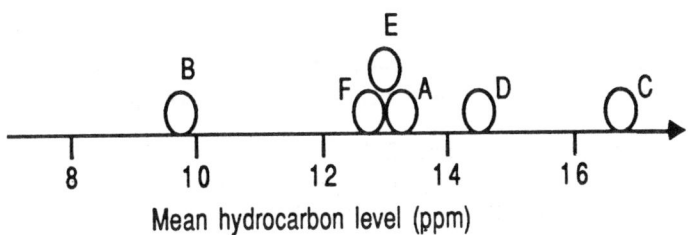

Fig. 2.1 — Variability between vessels.

The variability in Table 2.1 can be attributed to two causes. Firstly there is **variability from vessel to vessel** which is indicated by the means plotted in Fig. 2.1. The biochemist would argue that this variability from vessel to vessel results from the fact that the micro-organisms differ in their effectiveness. If he had put the **same** organism in each vessel most of this variability would not exist. Secondly there is **variability within each vessel**. This cannot be blamed on the micro-organisms but it can be attributed to a lack of precision in the measurements. Unfortunately the determination of hydrocarbon in water is not as accurate as the biochemist would wish. This is why he made two or even three measurements on each vessel, realizing that the mean of several observations is more reliable than a single observation. Why then does he not make 10 or 20 measurements on each vessel? Because the method of measurement is very time consuming and very expensive.

The means plotted in Fig. 2.1 suggest that organism B is clearly superior to the others, while organism C appears to be distinctly inferior. Of course, the means do not tell us anything about the errors of measurement in the data. When these errors are revealed in Fig. 2.2 we may be more reluctant to draw conclusions.

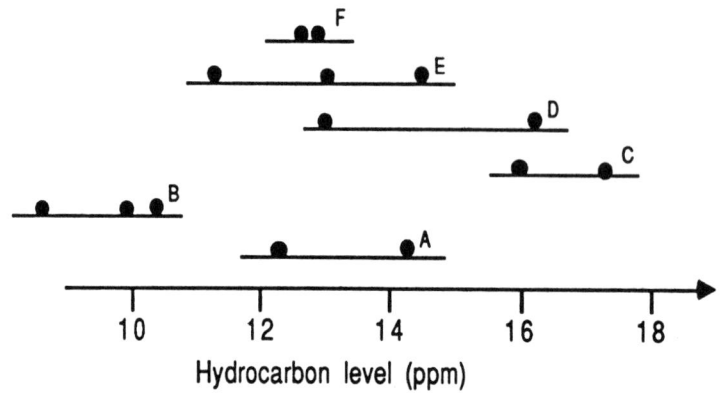

Fig. 2.2 — Variability between vessels and within vessels.

Exercise 2.2

The purpose of this question is to compare your common sense conclusions with those which will emerge from the sophisticated analysis of the data, which follows.

(a) Carefully examine Figs. 2.1 and 2.2. Are you prepared to conclude that organism B is superior to any or all of the other five?
(b) Calculate 95% confidence limits for the true hydrocarbon level in vessel B. Use the formula $\bar{x} \pm ts/\sqrt{n}$, with the value of t being obtained from Table ST1.
(c) What exactly to we mean by 'true hydrocarbon level' in part (b)?
(d) Repeat part (b) for the other five vessels.

You may be staggered by the enormous width of the confidence intervals you have just calculated. Obviously there is considerable overlap between these intervals as we can see in Fig. 2.3. This diagram certainly does not support the conclusion that organism B is superior to the other five. Our disappointment grows when we notice that two of the lower confidence limits are negative, although we realize that the hydrocarbon content of a vessel could not be less than zero.

I hope that the confidence intervals in Fig. 2.3 do not undermine the confidence that you built up while studying Chapter 1. I guess you feel that the intervals defy commonsense. I sympathize. However, I must insist that each confidence interval in Fig. 2.3 is perfectly reasonable from a statistical point of view. The width of an interval warns us of the enormous uncertainty that exists when we attempt to estimate a population mean using data from a very small sample. If we have only two observations in our sample then the value of t from Table ST1 is equal to 12.71. Thus

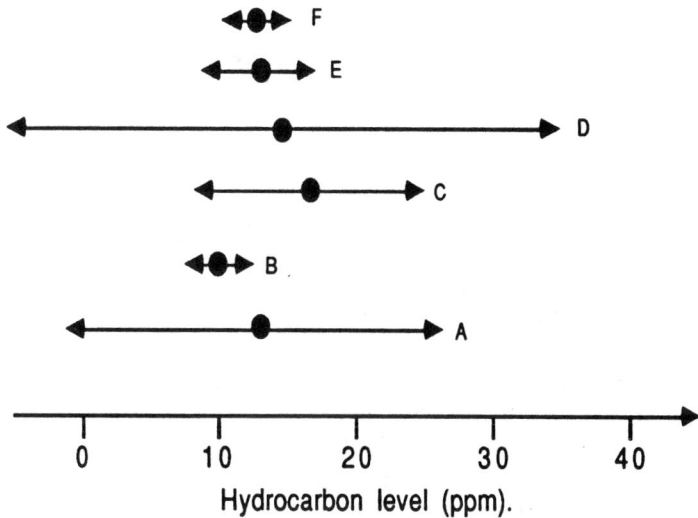

Fig. 2.3 — Confidence intervals for true hydrocarbon levels.

$t/\sqrt{n}$ will be equal to 8.99 and the confidence limits will be approximately 9 standard deviations from the mean. Thus the confidence interval is likely to be much wider than the spread of the data, when we have such a small sample.

If you accept that each confidence interval in Fig. 2.3 is quite reasonable, must you now abandon the conclusion reached in Exercise 2.2, that organism B is superior? It would be a great pity if you did, for your conclusion is also very reasonable. Later in this chapter we shall prove, beyond reasonable doubt, that organism B really is superior to the other five. Furthermore, we shall use a method of analysis which may well reflect the line of thought you adopted whilst doing Exercise 2.2(a).

Do not underestimate your powers of reasoning. The human mind is a very powerful analytical tool, especially when it is dealing with information presented in a pictorial form. While you pored over Fig. 2.3 you probably reasoned as follows:

(a) The variability within vessels tends to obscure any real difference between the organisms.
(b) Taking account of the variability within vessels, the mean for organism B is so much less than the other means, that I will conclude B is superior.

Does this correspond to your thinking? Perhaps you did not use expressions such as 'variability within vessels', but it is very likely that you intuitively took account of this variability. Herein lies the strength of your reasoning and the weakness of the confidence intervals in Fig. 2.3. You took an **overall view** of the data, whereas each confidence interval is based on only one small part of the data considered in isolation. Unfortunately, each mean and each standard deviation is based on very few data. Let us, therefore, modify our statistical technique so that it is based on an overall view of the variability.

2.3   COMBINING STANDARD DEVIATIONS

The first step in our data analysis is to combine the six standard deviations in Table 2.1 to obtain just one number. This **combined standard deviation** gives us an overall view of the variability within vessels. Common sense suggests that this is a reasonable calculation to carry out. The standard deviation for vessel A (1.485) is a measure of the variability within that vessel. This variability results from the lack of precision in the measurement of hydrocarbon content. Similarly the standard deviation for vessel B (0.929) is also due to the imperfection of the method of measurement. In fact all six standard deviations are due to testing error, and it is reasonable to combine the standard deviations to obtain one number which quantifies the variability of the test method.

To calculate the combined standard deviation we could simply add up the six standard deviations and divide by six. This average standard deviation would be equal to 1.2315. Being an average it must lie somewhere between the smallest standard deviation (0.141) and the largest (2.263), of course.

Unfortunately this simple average of the standard deviations could be misleading. It does not take account of the fact that our samples differ in size. In some vessels we have three measurements while in others we have only two. To allow for the different sample sizes we shall calculate a weighted average. There is a second reason why the simple average of standard deviations is unacceptable. It is not valid to add standard deviations. We must square each standard deviation to obtain a variance; then we add the variances and finally take the square root. Incorporating both of these ideas we obtain the following formula:

$$\text{Combined standard deviation} = \sqrt{\{\Sigma[(\text{df})(\text{SD}^2)]/\Sigma(\text{df})\}}$$

In the above formula Σ is a summation sign. It simply informs us that several items must be **added**. SD means standard deviation and df means degrees of freedom. You will recall that our calculator divides by n-1 when it is calculating a standard deviation. In Chapter 1 we referred to n-1 as the degrees of freedom. Thus a standard deviation based on two numbers has 1 degree of freedom and a standard deviation calculated from three numbers has 2 degrees of freedom, etc. Use of this formula with the standard deviations from Table 2.1 gives:

Combined standard deviation
$$= \sqrt{\{[(1)(1.485^2)+(2)(0.929^2)+(1)(0.919^2)+\ldots]/(1+2+1+1+2+1)\}}$$
$$= \sqrt{(15.735/8)}$$
$$\doteq 1.386$$

The combined standard deviation (1.386) is a kind of average and its value must lie between the smallest of the standard deviations (0.141) and the largest (2.263). In fact the combined standard deviation is quite close to the simple average (1.2315) that we calculated earlier. However, the two would not be so close with many sets of data. Now that we have obtained the combined standard deviation, there is one very important point that you should note.

> The combined standard deviation is a better estimate of the population standard deviation, because it has more degrees of freedom.

In this particular case the combined standard deviation has 8 degrees of freedom. This figure is obtained by adding together the degrees of freedom of all the standard deviations that went into the calculation. Thus our combined standard deviation is as good as a standard deviation calculated from nine measurements. The benefit of the extra degrees of freedom will be obvious when we use the combined standard deviation to calculate confidence intervals.

2.4   CONFIDENCE LIMITS BASED ON THE COMBINED STANDARD DEVIATION

To calculate confidence limits for the true hydrocarbon content of any vessel we need a mean and a standard deviation. However, these do not need to come from the same data. Thus, we can obtain confidence limits for vessel A using the mean of the two determinations from A (13.25) together with the combined standard deviation. We shall use the formula introduced in Chapter 1, $\bar{x} \pm ts/\sqrt{n}$, but we must keep in mind the following.

(a) When obtaining a value of t from the 95% column of Table ST1 we use the degrees of freedom of the combined standard deviation, i.e. 8 in this case.
(b) n is the number of observations from which the mean was calculated, i.e. two for vessel A.

95% confidence limits for the true hydrocarbon level in vessel A are calculated as follows:

$$\bar{x} \pm ts/\sqrt{n}$$
$$= 13.25 \pm (2.31)\,(1.386)/\sqrt{2}$$
$$= 13.25 \pm 2.26$$
$$= 10.99 \text{ to } 15.51 \text{ ppm}$$

Thus we can be 95% confident that the true hydrocarbon level for vessel A lies between 10.99 and 15.51 ppm. Clearly this interval is narrower than that obtained in Exercise 2.1. Similar calculations can be made for each of the other five vessels.

Exercise 2.3
(a) Using the combined standard deviation calculate 95% confidence limits for the true hydrocarbon content of vessel B.
(b) How would you answer the criticism that 'the calculation you carried out in part (a) is not valid because you used data from other vessels to draw conclusions about vessel B'?.

The confidence intervals for all six vessels are given in Table 2.2 and displayed in Fig. 2.4. These intervals are obviously much narrower than those calculated earlier and represented in Fig. 2.3. Use of the combined standard deviation has allowed us to make full use of the data and given confidence intervals which accord with common sense.

Table 2.2 — 95% confidence intervals using the combined standard deviation

Vessel	Mean	SD	t	n	Confidence interval	
A	13.25	1.386	2.31	2	13.25±2.26	(10.99 to 15.51)
B	9.63	1.386	2.31	3	9.63±1.85	(7.78 to 11.48)
C	16.75	1.386	2.31	2	16.75±2.26	(14.49 to 19.01)
D	14.60	1.386	2.31	2	14.60±2.26	(12.34 to 16.86)
E	13.00	1.386	2.31	3	13.00±1.85	(11.15 to 14.85)
F	12.80	1.386	2.31	2	12.80±2.26	(10.54 to 15.06)

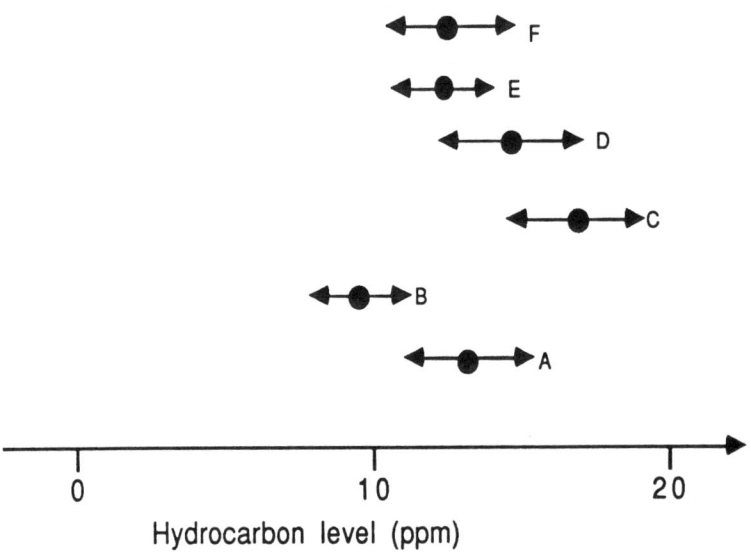

Fig. 2.4 — Confidence intervals from Table 2.2.

What conclusions can we now draw concerning the six micro-organisms? We can see in Fig. 2.4 that the confidence intervals for vessels B and C do not overlap at all. It seems reasonable, therefore, to conclude that organism B is superior to C. By a similar argument we can conclude that B is superior to D. However, the interval for B does overlap with the intervals for A, E and F. We may, therefore be tempted to conclude that B is not superior to these three organisms.

This line of reasoning is very conservative and it could well lead us to the conclusion that two organisms do not differ when in fact they do. Consider, for example, organisms B and E. The confidence intervals tell us that the mean for B **could** be as high as 11.48 while the mean for E **could** be as low as 11.15 and, therefore, the two means **could** be equal. Yes, they **could** be equal, but this is very unlikely. When we quote confidence limits we expect the population mean to be near the centre of the interval rather than at the edge. It is very unlikely, therefore, that **two** confidence intervals would each have their population means close to a confidence limit. To overcome this problem we shall adopt a slightly different approach, which is not so conservative. It will enable us to answer two questions simultaneously:
(a) Do two specified organisms differ in their ability to degrade the oil?
(b) How large is the difference?

2.5   CONFIDENCE LIMITS FOR THE DIFFERENCE BETWEEN TWO MEANS

If we conclude that one particular micro-organism is better than another, it is quite natural that we will wish to ask 'How much better?' Our best estimate of the superiority will be the difference between the two sample means. For organisms A and B the means are 13.25 and 9.63 ppm. The difference is 3.62 ppm. Thus we could conclude that B gives a reduction of 3.62 ppm more than A would give. However, as we noted in Chapter 1, any such estimate is almost certain to be wrong. Furthermore, the estimate alone gives no indication of its reliability or precision. Thus we prefer to quote two limits, rather than one estimate. What we need at this point are confidence limits for the difference we would find if we tested organisms A and B for ever more.

Confidence limits for the difference between two population means are given by:

$$(\bar{x}_1 - \bar{x}_2) \pm t\, s\sqrt{[(1/n_1) + (1/n_2)]}$$

where $\bar{x}_1$ is the larger of the two sample means, $\bar{x}_2$ is the smaller of the two sample means, n_1 is the number of observations in $\bar{x}_1$, n_2 is the number of observations in $\bar{x}_2$, s is an appropriate standard deviation and t is a value from Table ST1 with the same degrees of freedom as the standard deviation.

Let us concentrate on the micro-organisms in vessels A and B. We know from Table 2.1 that the mean hydrocarbon content for vessel A was 13.25 ppm and for B was 9.63 ppm. Thus $\bar{x}_1 = 13.25$, $\bar{x}_2 = 9.63$, $n_1 = 2$ and $n_2 = 3$. To obtain a suitable standard deviation we could combine the standard deviation for vessel A (1.485) with that for B (0.929). This would give us a combined standard deviation of 1.145 with 3 degrees of freedom. However, in this situation we can do better. We can use the combined standard deviation for all six vessels (1.386) which has 8 degrees of freedom. For 95% confidence the appropriate t value from Table ST1 is 2.31.

95% confidence limits for the difference between the true hydrocarbon levels of A and B are:

$$(13.25-9.63)\pm(2.31)\ (1.386)\sqrt{[(1/2)+(1/3)]}$$
$$=3.62\pm2.92$$
$$=0.70 \text{ to } 6.54$$

We can be 95% confident that the difference between the true hydrocarbon levels for vessels A and B lies between 0.70 and 6.54 ppm. The important point about this interval is that **it does not contain zero**. Thus we can conclude that organisms A and B really do differ. We have proved beyond reasonable doubt that organism B gives a greater reduction in hydrocarbon level than does A.

Exercise 2.4
(a) Calculate 95% confidence limits for the difference between the true hydrocarbon levels for vessels B and E.
(b) Can we reasonably conclude that the organisms in vessels B and E differ in their ability to reduce the hydrocarbon level in sea water?

2.6 MULTIPLE COMPARISONS

Using the procedure you have followed in Exercise 2.4, we could compare organism B with each of the other five. If we did so we would find that B was superior to each. 'Why stop at this point?' you might ask. Why, indeed. We could now compare F, which had the second lowest mean, with organisms A, C, D and E. Then we could compare A with C, D and E etc. We would need 15 confidence intervals to compare all possible pairs. By making this multitude of comparisons we would prove, beyond reasonable doubt, that certain organisms were superior to certain others. The conclusions we would reach by this approach are summarized in Fig. 2.5.

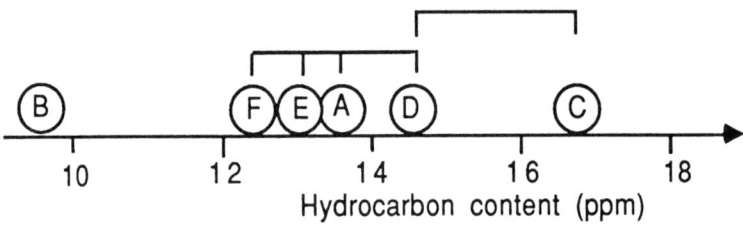

Fig. 2.5 — Comparison of six micro-organisms.

Fig. 2.5 is a simple dot plot, with each organism being represented by a dot or circle. You will see that some circles are joined by lines above. Those circles which are directly joined by a line are **not** significantly different. Thus we have proved that B is superior to all other organisms and F is superior to C, for example.

Fig. 2.5 is obviously very useful, once you have got used to the fact that it seems to tell you the opposite of what you want to know. It summarizes very succinctly the many comparisons we have made. Unfortunately, Fig. 2.5 could be misleading. To help you understand just how this diagram could lead us astray, I will need to focus upon the **risk** of drawing wrong conclusions. You may find the discussion rather tedious. If you do so, it might be wise to skip to the next section, as this argument will be repeated in a later chapter.

Unfortunately, whenever we use data from a sample to make a decision about a population, we run a risk of making a **wrong** decision. For example, we calculated confidence limits for the difference between the true hydrocarbon levels for vessels A and B which led us to the conclusion that B had a lower level than A. This conclusion could be wrong. It is **possible** that A has a lower level than B. The data strongly indicate that we have made the right decision, but we must acknowledge there is a possibility that we are wrong.

The use of statistical techniques does not eliminate the risk of making wrong decisions, but it does enable us to quantify the risk. Because we used 95% confidence when comparing vessels A and B, we can assert that the risk of being wrong is less than 5%. Whenever we find that our 95% confidence interval for the difference does not include zero, we conclude that there is a real difference, and the chance of being wrong is less than 5%. If we wanted to reduce this risk to less than 1% we would use a 99% confidence interval.

What I have said in the preceding paragraph applies to the making of **one** comparison. We have used the data in Table 2.1 to make 15 comparisons. In 8 of these 15 comparisons the confidence interval did **not** include zero and we therefore make 8 positive decisions; that organism B was superior to F, E, A, D, and C, while organism C was also inferior to F, E and A. The risk of error associated with each of these eight decisions is less than 5%. However, the **overall risk**, that is the risk of making at least one error, is much greater than the individual risks. It is because the risk is greater than it first appears to be that multiple comparisons are dangerous. Nevertheless, multiple comparisons are essential. The biochemist carrying out the research wishes to know which organisms are better than which other organisms. Furthermore, he cannot specify in advance which comparisons are important, because he knows little or nothing about the effectiveness of each organism.

2.7 SAMPLE SIZES

The biochemist who carried out the experiment to compare the six micro-organisms made two determinations on some vessels and three on others. He had planned to make four determinations on each, but the full programme of testing was never completed. Obviously, we could have compared the organisms with greater precision if we had had all the data that he originally planned to obtain. When the decision was made to terminate the testing, the biochemist was very concerned on two counts:

(a) Would the smaller samples be big enough?
(b) Would the differences in sample size invalidate the statistical analysis?

Common sense tells us that larger samples are better than smaller samples, and that it is wise to distribute our experimental effort equally between the samples. Close examination of our confidence interval formula:

$$(\bar{x}_1-\bar{x}_2)\pm t\,s\,\sqrt{[(1/n_1)+(1/n_2)]}$$

will support what common sense suggests. To obtain a narrower interval we need $[(1/n_1)+(1/n_2)]$ to be smaller. This can be achieved by making n_1 and/or n_2 larger. Furthermore, for a given total sample size, this term will be smallest when n_1 is equal to n_2. Table 2.3 illustrates this point for a total sample size of 8.

Table 2.3 — Equal sample sizes give a narrower interval

n_1+n_2	n_1	n_2	$\sqrt{(1/n_1+1/n_2)}$
8	1	7	1.069
8	2	6	0.816
8	3	5	0.730
8	4	4	0.707

Suppose that you wished to carry out an experiment to compare two treatments. The message from Table 2.3 is that you should share your total effort so that you obtain equal numbers of results with each treatment. We are still left with the question, 'How many results do you need in total?' As we saw in Chapter 1, when dealing with only one treatment, the size of sample required depends on two factors; how accurately you wish to assess the effect and how variable is the situation. These factors are clearly seen in the sample size formula we used in Chapter 1:

$$n=(ts/c)^2$$

where s is a suitable standard deviation and n is the sample size needed to obtain a confidence interval of $\pm c$. When we are comparing two treatments we need two samples, of course. If we wish to have two equally large samples then the required size is given by the following formula.

The sample sizes needed to estimate the difference between two population means, to within $\pm c$, are:

$$n_1=n_2=2(ts/c)^2$$

where n_1 and n_2 are the sample sizes,
  s is a suitable standard deviation,
  t is taken from Table ST1, with the degrees of freedom appropriate to the standard deviation, and
  $\pm c$ is the desired width of confidence interval.

We cannot use the above formula unless we have a suitable standard deviation. This standard deviation should be an estimate of the variability that we would find in the data if we had an unlimited number of measurements with the **same** treatment. In the comparison of micro-organisms, for example, the standard deviation should reflect the variability among determinations made on the same vessel. Thus the combined standard deviation is acceptable for this purpose. Let us use the combined standard deviation, 1.386, to estimate the size of sample needed to obtain a confidence interval with a width of ± 1.0 ppm:

$$n_1 = n_2 = 2(ts/c)^2$$
$$= 2[(2.31)(1.386)/1]^2$$
$$= 20.5$$

Obviously sample sizes must be whole numbers so we round this result to 21. Thus we would need 21 determinations on each vessel in order to obtain a confidence interval which was no wider than ± 1.0 ppm.

Exercise 2.5

The effectiveness of two oil additives has been assessed by running 12 identical engines for 100 hours. Four engines were allocated to each additive and four were given standard oil. Cylinder wear for each engine is measured. The results are summarized in Table 2.4.

Table 2.4 — Cylinder wear of 12 engines

Additive	None	X	Y
Mean	16.4	14.3	13.7
SD	2.067	3.270	2.815

(a) Calculate the combined standard deviation for these data.
(b) Calculate the size of samples needed in a future comparison of additives P and Q if we wish to assess the difference between the two means to within ± 2.0 thou.
(c) Suppose the three standard deviations in Table 2.4 had been 0.067, 3.270 and 2.815. How would you modify your calculations in parts (a) and (b)?

2.8 ASSUMPTIONS

In the final section of Chapter 1 I pointed out the dangers associated with the use of statistical methods. The more powerful techniques introduced in this chapter are just as dangerous, because they share the same assumptions that we discussed in Chapter 1. You will recall that these assumptions are built into the statistical tables. We have used Table ST1 while calculating confidence limits in this chapter, just as we did in Chapter 1. The two assumptions associated with Table ST1 are:

(a) that the sample was selected at random from the population;
(b) that the population has a normal distribution.

In this chapter we have extended the confidence interval technique to cover **two** samples. We therefore need to add a third assumption to our list:

(c) that the two populations are equally variable.

This additional assumption was hinted at in Exercise 2.5(c) when I pointed out how unwise it would be to combine two standard deviations which were very different. If the larger standard deviation is many times greater than the smaller, it is hard to believe that the two samples came from populations which are equally variable. In a later chapter I shall introduce a simple test that we can use to check the compatibility of standard deviations before we combine them.

2.9 A SUMMARY OF THE IMPORTANT POINTS

(1) Whenever we compare sample means we must take account of the variability in the data. To do so we use the standard deviations of the samples.
(2) The usefulness of a standard deviation depends very much on its degrees of freedom.
(3) Combining two or more standard deviations is one way of increasing our degrees of freedom. We use the formula:

$$\text{combined standard deviation} = \sqrt{\{\Sigma[(\text{df})\,(\text{SD}^2)]/[\Sigma(\text{df})]\}}$$

The degrees of freedom of the combined standard deviation are equal to Σ (df).
(4) It is only valid to combine standard deviations if they are measures of the same source of variation (i.e. if they are due to the same cause).
(5) Confidence limits for the difference between two population means are given by:

$$(\bar{x}_1 - \bar{x}_2) \pm t\,s[(1/n_1) + (1/n_2)]$$

(6) If a confidence interval for the difference between two means **does not include zero** we can reasonably conclude that the population means do differ.
(7) In the comparison of **many** treatments, two at a time, we run a greater risk of 'finding' a difference that does not exist.
(8) When comparing two treatments our confidence interval for the difference will be narrowest if we have equal numbers of determinations on each.
(9) The sample sizes needed to estimate the difference between two population means to within $\pm c$ are given by:

$$n_1 = n_2 = 2(t\,s/c)^2$$

(10) When we use a confidence interval to compare two treatments we invoke the additional assumptions that the two populations are equally variable.

2.10 ADDITIONAL EXERCISES

Exercise 2.6
An agricultural scientist wishes to compare two varieties of wheat. He prepares a

large field then subdivides this into ten plots of equal size. His intention is to assign five plots to each variety randomly, but his instructions are misunderstood and six plots are planted with variety A while only four are planted with variety B. The yields in tons per hectare of the ten plots are given in Table 2.5.

Table 2.5 — Wheat yields

Plot	1	2	3	4	5	6	7	8	9	10
Variety	A	A	B	A	B	B	A	A	B	A
Yield	3.9	4.6	3.0	4.1	3.5	3.3	3.9	3.4	2.2	2.6

(a) Draw dot plots to compare the yields of the two varieties.
(b) Calculate the mean yield and the standard deviation for each variety.
(c) Use the two standard deviations from part (b) to calculate a combined standard deviation.
(d) Think about the meaning of the standard deviations you have calculated in (b) and (c). Would you expect to get larger or smaller standard deviations if:
　(i) some plots had been given more fertilizer than other plots?
　(ii) some plots had better drainage than other plots?
(e) Calculate 95% confidence limits for the difference in mean yield between these two varieties of wheat.
(f) Can we reasonably conclude that variety A gives a higher yield than variety B?
(g) Can we reasonably conclude that variety B gives a higher yield than variety A?
(h) How many plots of land of this size would be needed in order to estimate the difference between the two varieties to within ±0.5 tons per hectare?

Exercise 2.7
The agricultural scientist in Exercise 2.6 is surprised that variety A did not prove to be superior to B and he re-examines the data for signs of abnormality. He notices that plots 9 and 10 gave a much lower yield than the other plots. His assistant, who carried out the experiment, reveals that the ten plots were parallel strips and that plots 9 and 10 may have been contaminated by industrial waste deposited in the adjacent field.

(a) Remove the data for plots 9 and 10 and then recalculate the means and standard deviations for the two varieties.
(b) Calculate a combined standard deviation and 95% confidence limits for the difference in yield between the two varieties.
(c) Can we reasonably conclude that variety A is superior to B?
(d) Do you consider that the agricultural researcher is justified in re-analysing his data and drawing a different conclusion?

2.11 WORKED SOLUTIONS

Solution to Exercise 2.1

(a) There could be many reasons why these two determinations differ. I have thought of the following, but you may have listed others.

(i) There could be a real difference between the hydrocarbon levels in vessels B and C. This difference could be due to organism B's being more effective than A, or a difference in hydrocarbon levels at the outset, or different conditions in the two tanks (e.g. temperature).

(ii) Measurement error. The test method for determining the hydrocarbon content is very time consuming and thus expensive. If we knew the repeatability of the test method we would be able to assess whether or not such a large difference could be attributed to measurement error. (See Volume 3 of this series, *Statistics for Analytical Chemists.*)

(iii) Sampling error. Perhaps the hydrocarbon is not dispersed homogeneously throughout the vessel, in which case the analytical sample might not be representative of the content of the vessel.

(b) As these two determinations were made on samples from the same vessel we can rule out the first of the three explanations from part (a). However, the difference could be due to either sampling error or measurement error, or both.

(c) If all 14 determinations were equal we could conclude that there was no measurement error, no sampling error and no differences between the six micro-organisms. However, it would seem very odd that any sensible experiment would yield such results.

Solution to Exercise 2.2

(a) It is impossible for me to know what conclusions you have reached. However, I anticipate that many readers will have concluded that organism B is superior to each of the other five. Perhaps the most convincing evidence to support such a conclusion is the fact that every one of the three measurements on vessel B is less than any measurement on any other vessel.

(b) For vessel B, mean=9.63 and SD=0.929. From Table ST1 t=4.30 with 2 degrees of freedom. Therefore the confidence limits are

$$9.63\pm4.30\ (0.929)/\sqrt{3}$$
$$=9.63\pm2.31$$
$$=7.32 \text{ to } 11.94 \text{ ppm}$$

We can be 95% confident that the true hydrocarbon level of vessel B was between 7.32 and 11.94 ppm.

(c) The 'true hydrocarbon level' can be defined as the mean determination we would obtain if we continued making measurements for ever more. Other definitions are possible but all should indicate that the measurements actually made are a sample from the population of measurements which could have been made.

(d) The confidence intervals for all six vessels are given in Table 2.6.

Solution to Exercise 2.3

(a) Sample mean for vessel B=9.36 ppm $\bar{x}$=9.63
   Combined standard deviation=1.386 s=1.386
   Number of measurements made on vessel B=3 n=3

Table 2.6 — 95% confidence intervals for true hydrocarbon level

Vessel	Mean	SD	t	Confidence interval	
A	13.25	1.485	12.71	13.25±13.35	(−0.10 to 26.60)
B	9.63	0.929	4.30	9.63± 2.31	(7.32 to 11.94)
C	16.75	0.919	12.71	16.75± 8.26	(8.49 to 25.01)
D	14.60	2.263	12.71	14.60±20.34	(−5.74 to 34.94)
E	13.00	1.652	4.30	13.00± 4.10	(8.90 to 17.10)
F	12.80	0.141	12.71	12.80± 1.27	(11.53 to 14.07)

Value from Table ST1=2.31 $t=2.31$

95% confidence limits for the true hydrocarbon level in vessel B are given by

$$\bar{x}\pm t\,s/\sqrt{n}$$
$$=9.63\pm(2.31)\,(1.386)/\sqrt{3}$$
$$=9.63\pm1.85$$
$$=7.78 \text{ to } 11.48$$

We can be 95% confident that the true hydrocarbon content of vessel B lies between 7.78 and 11.48 ppm.

(b) In the formula $\bar{x}\pm t\,s/\sqrt{n}$, we are using the sample mean $\bar{x}$ to estimate the population mean. Similarly, we are using the calculated standard deviation s to estimate the population standard deviation. If we knew the population standard deviation, we would use it. However, as we do not know the population standard deviation, it makes sense to use the best available estimate. Thus, before we used the combined standard deviation in part (a) we should have asked, 'Is it the best estimate of the population standard deviation?'

Perhaps you will understand the question better if it is re-worded in more practical terms: 'Is the combined standard deviation the best available estimate of the variability we would find if we made many determinations on samples from vessel B?' In my opinion it is reasonable to answer 'yes'. I say this because, whichever vessel we examine, the variability that we find is due to the same cause . . . testing error (and perhaps sampling error). (Note, however, that I would not hold this opinion if one of the six standard deviations were much greater than the other five. We shall return to this point in Chapter 4, when we discuss the assumptions underlying statistical techniques.)

Solution to Exercise 2.4

(a) From Table 2.1 we obtain the mean hydrocarbon levels for vessels B and E.

$$\bar{x}_1=13.00 \qquad \bar{x}_2=9.63 \qquad n_1=3 \qquad n_2=3$$

The combined standard deviation based on all six vessels is equal to 1.386, with 8 degrees of freedom. The t value from Table ST1, for 95% confidence, with 8 degrees of freedom is 2.31.

95% confidence limits are given by

$$(\bar{x}_1-\bar{x}_2)\pm t\,s\sqrt{[(1/n_1)+(1/n_2)]}$$
$$=(13.00-9.63)\pm(2.31)\,(1.386)\,\sqrt{[(1/3)+(1/3)]}$$
$$=3.37\pm2.61$$
$$=0.76 \text{ to } 5.98$$

Thus we can be 95% confident that the difference between the true hydrocarbon levels for vessels B and E lies between 0.76 and 5.98 ppm.

(b) As the confidence interval does not include zero we can be 95% confident that there is a real difference between the hydrocarbon levels in vessels B and E. Thus we can reasonably conclude that the organism in B is more effective than that in E.

Solution to Exercise 2.5

(a) Combined $SD=\sqrt{\{\Sigma[(df)(SD)^2]/[\Sigma(df)]\}}$
$$=\sqrt{\{[3(2.067)^2+3(3.270)^2+3(2.815)^2]/(3+3+3)\}}$$
$$=2.762$$

(b) $n_1=n_2=2(t\,s/c)^2$
$$=2[(2.26)\,(2.762)/2.0]^2$$
$$=19.5$$

Thus we would need 20 engines, run for 100 hours, with each additive.

(c) In this chapter we have combined two or more sample standard deviations in order to obtain a better estimate of the population standard deviation. However, this practice only makes sense if the samples can be regarded as coming from the **same** population. It is hard to believe that a sample of four items with a standard deviation of 0.067 came from the same population as the other samples, which had standard deviations of 3.270 and 2.815. Perhaps the introduction of either additive increases the **variability** in wear, in addition to any effect on the mean wear. Whatever the explanation, it seems unwise to combine all three standard deviations.

It would be better to combine only the 3.270 and the 2.815, and to use the result to estimate the required sample size. When the experiment is complete, we would be very interested to see whether the standard deviations of the P and Q samples are similar to those given by the X and Y samples.

Solution to Exercise 2.6

(a)

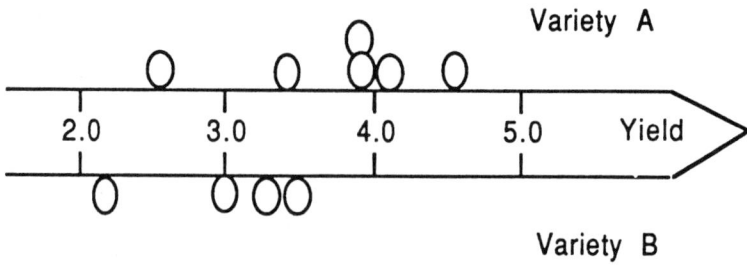

Fig. 2.6 — Comparison of two varieties of wheat.

(b) variety A: mean=3.75   SD=0.683 37
     variety B: mean=3.00   SD=0.571 55
(c) combined standard deviation=$\sqrt{\{\Sigma[(df)(SD^2)]/[\Sigma(df)]\}}$
$$=\sqrt{\{[(5)(0.683\,37^2)+(3)(0.571\,55^2)]/(5+3)\}}$$
$$=0.643\,72$$

(d) The standard deviations quantify the variation in yield from plot to plot. Probably the main cause of this variation in yield is the variability of the soil.
   (i) If some plots had been given more fertilizer than other plots this might well exaggerate the effect of soil variability and increase the standard deviations.
   (ii) If some plots had better drainage than other plots the standard deviations would probably increase.
        In agricultural experiments the researcher strives to create comparable conditions for the different varieties by using many small plots of land rather than using one large plot for each variety. This practice is based on the assumption that smaller plots are more homogeneous.

(e) $\bar{x}_1=3.75$ $\bar{x}_2=3.00$ $n_1=6$ $n_2=4$ $s=0.643\,72$ $t=2.31$
     95% confidence limits for the difference between the mean yields of the two varieties are given by
$$(\bar{x}_1-\bar{x}_2)\pm t\,s\sqrt{[(1/n_1)+(1/n_2)]}$$
$$=(3.75-3.00)\pm(2.31)(0.643\,72)\sqrt{[(1/6)+(1/4)]}$$
$$=0.75\pm0.96$$
$$=-0.21\text{ to }1.71$$
     Thus we can be 95% confident that the difference in yield between the two varieties lies between −0.21 and 1.71 tons per hectare.

(f) A simple comparison of the two sample means (3.75 and 3.00) leads us to the conclusion that variety A is superior to B by 0.75 tons per hectare. However, such a comparison takes no account of the variability within the data or the limited size of the experiment. A comparison based on a confidence interval is more reliable. The upper confidence limit implies that variety A might give as much as 1.71 tons per hectare more than B. The lower confidence limit, being negative, implies that variety B might be superior to A by as much as 0.21 tons per hectare. We have therefore, failed to prove beyond reasonable doubt that variety A is superior to B.

(g) We certainly **cannot** conclude that variety B is superior to A. The simple comparison of means suggested that A was superior; then the confidence interval revealed that the evidence was inconclusive, but this does not imply that B is superior.

(h) $n_1=n_2=2(t\,s/c)^2$
$$=2[(2.31)(0.643\,72)/0.5]^2$$
$$=17.7$$
     Thus we would need 18 such plots for each variety.

Solution to Exercise 2.7
(a) After removal of the data from plots 9 and 10 we obtain:
     variety A: mean=3.98   SD=0.432 43
     variety B: mean=3.27   SD=0.251 66

(b) combined standard deviation$=\sqrt{\{\Sigma[(df)\,(SD^2)]/[\Sigma(df)]\}}$
$$=\sqrt{\{[(4)(0.432\,43^2)+(2)\,(0.251\,66^2)]/(4+2)\}}$$
$$=0.381\,80$$

$\bar{x}_1=3.98$ $\bar{x}_2=3.27$ $n_1=5$ $n_2=3$ $s=0.381\,80$ $t=2.45$

95% confidence limits for the difference in yield between the two varieties are given by

$$(\bar{x}_1-\bar{x}_2)\pm t\,s\sqrt{[(1/n_1)+(1/n_2)]}$$
$$=(3.98-3.27)\pm(2.45)\,(0.381\,80)\,\sqrt{[(1/5)+(1/3)]}$$
$$=0.71\pm0.68$$
$$=0.03\ to\ 1.39$$

Thus we can be 95% confident that the difference in yield between the two varieties lies between 0.03 and 1.39 tons per hectare.

(c) Because the confidence interval does not include zero we can reasonably conclude that variety A gives greater yield than variety B.

(d) It is possible to use statistical techniques to confuse or to deceive people. If a scientist excludes part of his data and fails to report that he has done so, he may well deceive readers of his report. Regardless of his motives this is bad practice. However, if a scientist rejects part of his data and explains in his report why he has done so, then the reader will not be deceived and can judge whether or not the rejection was justified. With this set of data the proximity of the industrial waste tip could be considered sufficient justification for the rejection of the data from plots 9 and 10.

2.12 DETAILED OBJECTIVES FOR THIS CHAPTER

Now that you have studied this chapter and attempted to relate its content to your existing knowledge, you should be able to do the following.

(1) Explain the meaning of the following terms and use them appropriately in suitable contexts:
(a) combined standard deviation;
(b) confidence interval for the difference between two population means;
(c) 'we may reasonably conclude that the population means differ'.
(2) Calculate a combined standard deviation using the formula:
$$\sqrt{\{\Sigma[(df)(SD^2)]/[\Sigma(df)]\}}$$
(3) Calculate the degrees of freedom of a combined standard deviation using $\Sigma(df)$.
(4) Calculate confidence limits for the difference between two population means, using the formula:
$$(\bar{x}_1-\bar{x}_2)\pm t\,s\sqrt{[(1/n_1)+(1/n_2)]}$$
(5) Use Table ST1 to obtain a value of t for the calculation of confidence limits.
(6) Draw a conclusion about whether or not two population means differ, based on confidence limits for the difference between the means.
(7) Explain the risks associated with multiple comparisons in which several means are compared two at a time.
(8) Calculate the sample sizes needed to estimate the difference between two population means to within $\pm c$, using the formula:
$$n_1=n_2=2(t\,s/c)^2$$

(9) Distinguish between situations in which it would be reasonable to combine
standard deviations and those in which it would not.

2.13 SELF-TEST

Dr Protus is conducting an experiment to compare two alternative pig foods. 15 pigs
of the same age and each from a different litter are assigned at random to either feed
A or feed B. The weights of the pigs at the end of the 30 day trial are given in Table
2.7.

Table 2.7 — Weights of 15 pigs on two diets

Feed	Weight of pigs (kg)	Mean	SD
A	22.4, 26.6, 18.8, 20.2, 19.6, 24.5 28.0, 19.9	22.50	3.4897
B	27.6, 30.0, 24.5, 30.2, 22.6, 28.7, 24.0	26.80	3.07734

Questions (1)–(6) refer to the above data. As these questions are somewhat
interdependent you might wish to check your answer to question (1) before
attempting question (2), etc.
(1) What is the combined standard deviation for the data in Table 2.7?
    (a) 3.274
    (b) 3.283
    (c) 3.289
    (d) 3.305
(2) How many degrees of freedom does the combined standard deviation have?
    (a) 12
    (b) 13
    (c) 14
    (d) 15
(3) Which of the following is a 95% confidence interval for the population mean
    weight of pigs on diet A, based on the combined standard deviation?
    (a) 19.59 to 25.41 kg
    (b) 19.74 to 25.26 kg
    (c) 19.98 to 25.02 kg
    (d) 20.52 to 24.48 kg
(4) Which of the following is a 95% confidence interval for the difference in mean
    weights of the two feeds?
    (a) −0.38 to 8.98 kg
    (b) 0.61 to 7.99 kg
    (c) 2.39 to 6.21 kg
    (d) 2.46 to 6.14 kg
(5) Can we reasonably conclude that feed B gives a greater weight than feed A on
    average?

   (a) No, becuase some pigs on diet B weighed less than some on diet A at the end of the trial.

   (b) No, because the confidence interval in question (3) would overlap with a similar confidence interval for diet B.

   (c) Yes, because the confidence interval in question (3) does not include 26.8.

   (d) Yes, because the confidence interval in question (4) does not include zero.

(6) Suppose that the standard deviation of the eight weights with feed A had been much greater than the standard deviation of the seven weights with feed B. Which of the following is the most likely explanation?

   (a) The pigs assigned to feed A differed much more than those assigned to feed B.

   (b) A larger sample is expected to have a larger standard deviation than a smaller sample.

   (c) Whenever we take two samples the one with the larger mean is very likely to have the smaller standard deviation.

   (d) Feed A is manufactured more consistently than feed B.

(7) Dr Protus wishes to compare feeds X and Y, using similar pigs to those which gave the weights in Table 2.7. What sample sizes will he need if he wishes to obtain a confidence interval of ± 2 kg for the difference between the means?

   (a) 8 pigs fed on X and 8 on Y;

   (b) 13 pigs fed on X and 13 on Y;

   (c) 26 pigs fed on X and 26 fed on Y;

   (d) 51 pigs fed on X and 51 fed on Y.

(8) Which of the following statements is true?

   (a) We combine two or more standard deviations in order to obtain a better estimate of the population standard deviation.

   (b) We combine two or more standard deviations in order to obtain a wider confidence interval.

   (c) We should only combine standard deviations if they are the result of **different** causes.

   (d) We should only combine standard deviations if they have the **same** degrees of freedom.

(9) A medical researcher wishes to compare three anaesthetics (A, B and C), which are used during major surgery. Each of 50 hip replacement patients is assigned at random to one of the three anaesthetics and the total blood loss is recorded for each patient. Which is the best way to compare the mean blood loss for anaesthetics A and B?

   (a) Obtain a confidence interval using $\bar{x}_A$, $\bar{x}_B$ and the combined standard deviation; then ask 'Does the interval include zero?'

   (b) Obtain two confidence intervals using $\bar{x}_A$, $\bar{x}_B$, SD of A and SD of B; then ask 'Do the two intervals overlap with each other?'

   (c) Obtain two confidence intervals using $\bar{x}_A$, $\bar{x}_B$ and the combined standard deviation; then ask 'Do the two intervals overlap with each other?'

   (d) Obtain two confidence intervals using $\bar{x}_A$, $\bar{x}_B$ and the combined standard deviation; then ask 'Do either of the two intervals include zero?'

(10) In the situation described in question (9) it would **not** be reasonable to combine the three standard deviations if:

(a) One of the three samples of patients had a much higher mean blood loss than the other two.
(b) One of the three samples of patients had a much greater standard deviation than the other two.
(c) One of the three samples of patients was larger than the other two.
(d) None of the above.

2.14 ANSWERS TO SELF-TEST QUESTIONS

(1) (d)
(2) (b)
(3) (c)
(4) (b)
(5) (d)
(6) (a)
(7) (c)
(8) (a)
(9) (a)
(10) (b)

3

Controlling the process

3.1 INTRODUCTION

Almost all experiments can be put into one of three categories, depending on their purpose. The simplest experiments are those designed to **assess** quality or quantity. These were explored in Chapter 1. The second type of experiment is used to make **comparisons**, as we saw in Chapter 2. The third type of experiment is concerned with **relationships** between variables. In this chapter we shall explore statistical methods which have proved useful for scientists investigating relationships between process variables. The chapter has three aims:

(1) The primary aim of this chapter is to help you to master the technique of simple linear regression analysis. This is used to obtain the equation of the best straight line and to decide whether or not a relationship exists.
(2) Predictions or decisions based on regression analysis are more useful if accompanied by confidence limits. A second aim of this chapter is to help you to obtain and interpret appropriate confidence intervals.
(3) As in Chapters 1 and 2 we shall discuss the assumptions underlying any techniques which are introduced. Thus, the third aim is to familiarize you with these assumptions and the methods by which they can be checked.

A characteristic of the chemical industry is the extreme complexity of its production processes. From the day that a plant is commissioned to the time when it ceases to produce, we are engaged in a continuous struggle to gain a better understanding of its operation. We hope that the increase in understanding will help us to achieve better performance, in terms of reduced cost or increased quality–quantity.

The data we gather during routine operation and the data from special investigations can help us to increase our understanding of the process. In the analysis of this data we are mainly concerned with relationships between variables. On the one hand we have measures of the quality and quantity of the product. These cannot be controlled directly. On the other hand we have the operating conditions and the raw materials, which can be adjusted within certain limits. In this chapter we shall study a technique which helps us to explore the relationships between the two sets of variables.

3.2 CONTROLLING MOISTURE CONTENT

Peter Hubard is a chemist working in the Research and Development Department of Chemspec plc. He has particular responsibility for the rubber chemicals manufactured by Chemspec, which are sold mainly to tyre manufacturers. Recent complaints from a customer concerning excessive moisture content in Vulcatuf 587 have led to Dr Hubard's investigating the performance of the moving belt dryer through which this product passes.

Vulcatuf leaving the reactor has a moisture content of approximately 20%. This should be reduced to less than 4% as the product passes through the dryer. In order to achieve this target the operator could vary the belt speed, the belt temperature, the air temperature and the fan speed. However, the freedom of the operator is somewhat curtailed because values are specified for the belt temperature, the air temperature and the fan speed. In practice, therefore, he adjusts the belt speed to obtain the required moisture content.

Decreasing the speed will reduce the moisture in the final product. Unfortunately, it will also reduce the throughput, so overdrying can be an expensive refinement. There are conflicting opinions at Chemspec on how a compromise between quality and quantity should be achieved. Dr Hubard feels that the discussion of optimum belt speed would be placed on a firmer foundation if he could precisely quantify the two important relationships:

(a) between belt speed and moisture content;
(b) between belt speed and throughput.

For ten consecutive batches of Vulcatuf he ensures that the fan speed, air temperature and belt speed are held constant. The throughput, moisture content and belt speed are recorded for each batch, with the results shown in Table 3.1.

Dr Hubard has no doubt whatsoever that the moisture content and the throughput would both increase if the belt speed were increased. Thus it is not the purpose of the data analysis to prove that these relationships exist. The purpose is to quantify the relationships that are known to exist and then to predict the most suitable belt speed. Before we carry out a more formal analysis, I should like you to examine the data in Table 3.1 and to make several estimates.

Exercise 3.1

(a) What belt speed would you recommend in order to obtain a moisture content of 3.5% on average?
(b) What is the highest belt speed that will give 95% chance of obtaining a moisture content of 3.5% or less?
(c) If the operator used the belt speed you recommended in part (a), what throughput would you expect him to achieve?

Table 3.1 — Ten batches of Vulcatuf 587

Batch	Belt speed (cm/s)	Moisture content (%)	Throughput (kg/h)
627	2.9	2.8	159
628	3.2	3.3	177
629	2.6	2.3	151
630	3.4	3.7	193
631	3.1	3.4	167
632	3.1	3.7	171
633	3.4	4.1	191
634	2.6	2.8	148
635	3.3	3.9	186
636	2.8	3.4	154

The worked solution to Exercise 3.1 focuses on two scatter diagrams, with a straight line drawn on each, so as to fit the data as closely as possible. We often refer to such a line as the 'best straight line' or the 'regression line'. The equation of the best straight line is known as the 'regression equation'. This equation will have the form:

$$\text{moisture content} = a + b(\text{belt speed})$$

To fit the best straight line to a set of data we simply calculate values for a, which is known as the intercept, and b, which is known as the slope. The equation becomes $y = a + bx$, of we follow the accepted convention of representing the two variables by the letters x and y. In fact there are **several** conventions associated with regression lines and regression equations. If we do not rigidly adhere to these conventions, there is considerable scope for confusion. The conventions are:

(a) The variable on the right-hand side of the equation is known as the **independent variable**, or the **predictor variable**, or the **control variable**.
(b) The variable on the left-hand side of the equation is known as the **dependent variable**, or the **response variable**, or simply the **response**.
(c) The independent variable is represented by x and the dependent variable by y.
(d) In a scatter diagram the independent variable (x) is on the horizontal axis and the dependent variable (y) is on the vertical axis.
(e) When we type pairs of numbers into our pocket calculator to obtain the slope (b) and intercept (a) of the regression equation, the value of x precedes the corresponding value of y.

It will be clear to anyone who intends to observe these conventions that the first task in any regression analysis is to decide which of the two variables is the independent variable. In our discussion of the relationship between belt speed and moisture content we will regard belt speed as the independent variable and moisture content as the response. This is not a difficult decision to make as there is obviously a cause–effect relationship; changes in belt speed (x) **cause** changes in moisture content (y).

When we are investigating chemical processes the independent variables are those which can be controlled directly, such as feed rate, temperature, pressure, ratio of ingredients and speed of agitation. The dependent variables will relate to the quantity and quality of the end product, such as yield, impurity, throughput, taste and appearance. In many situations, however, it would be meaningless to speak of cause and effect, and hence different criteria must be used. We shall discuss this point again later in the chapter.

3.3   FITTING THE BEST STRAIGHT LINE

In order to study the remainder of this chapter effectively you will need a suitable calculator. Unfortunately, it is difficult for me to advise you on what type of calculator you should buy or borrow. On my desk is a Casio fx-180P scientific calculator which I bought in 1982 for £19.95. If you have the same type of calculator you are well equipped for studying this chapter. However, if you have not got one, there is no point in rushing out to buy one, for it is no longer obtainable. Furthermore, the new model which superseded the fx-180P will itself be superseded by the time you study this chapter. Clearly, it is impossible for me to advise you which type of calculator to obtain.

Perhaps you should take this chapter to your calculator supplier and ask him or her to fit the best straight line to the moisture and speed data in Table 3.1. Any calculator which gives the equation $y=-1.545+1.607x$ will be suitable. (If your supplier suggests that the equation of the best straight line is $y=1.474+0.469x$, he has probably interchanged the x and y numbers.)

The intercept (-1.565) and the slope (1.607) of the best straight line are obtained very easily on a suitable calculator. Simply select the appropriate mode, clear the memories, then type in the data. All the necessary calculations are done automatically and the following results appear in the display:

(a)   the mean belt speed$=3.04$;
(b)   the mean moisture content$=3.34$;
(c)   the standard deviation of the belt speed$=0.302\,581$;
(d)   the standard deviation of the moisture content$=0.560\,159$;
(e)   the intercept $a=-1.544\,66$;
(f)   the slope $b=1.606\,796$;
(g)   the correlation coefficient$=0.867\,945$.

The slope of the best straight line tells us how much the dependent variable will change as we increase the independent variable by one unit. Thus we can expect an increase in moisture content of approximately 1.6% if we increase the belt speed by 1cm/s. The intercept of the best straight line is the value of the dependent variable when the value of the independent variable is equal to zero. Thus the results of our calculation imply that a belt speed of zero would give a moisture content of -1.54%.

Obviously a negative moisture content is not possible. How can Dr Hubard have confidence in an equation which gives such a silly prediction? You are asked to consider this, and other questions, in Exercise 3.2, which refers to Figs. 3.1 and 3.2.

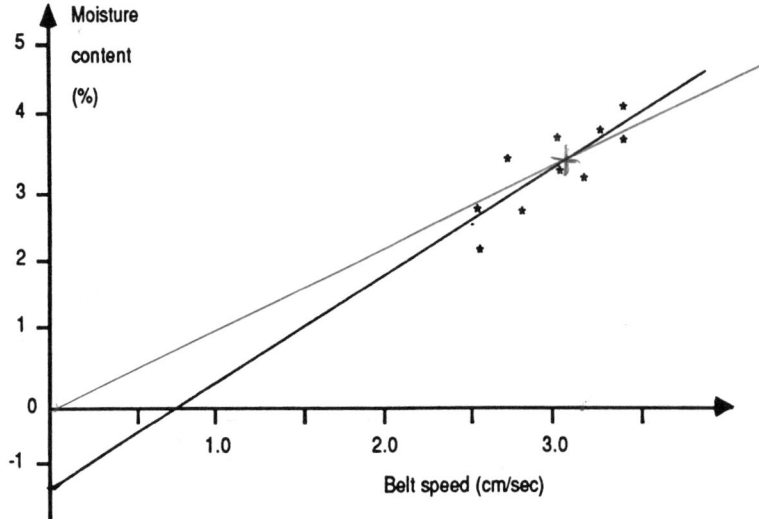

Fig. 3.1 — The best straight line.

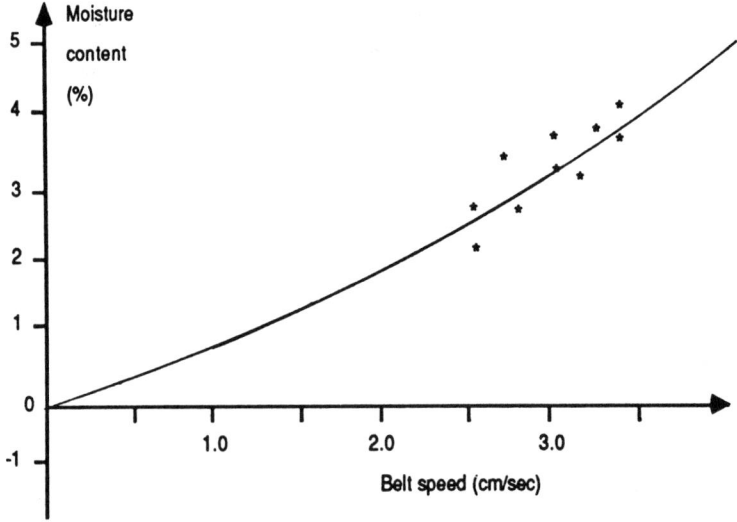

Fig. 3.2 — Is the curve better than the best straight line?

Exercise 3.2

(a) Mark a cross on Fig. 3.1 at the centre of the data, using the mean belt speed of 3.04 cm/s and the mean moisture content of 3.34%. The point you have marked is known as the **centroid**. Note that the best straight line passes through the centroid. For any set of data the regression line will pass through the centroid.

(b) Draw on Fig. 3.1 a straight line which has the equation $y = 1.103x$. This line is known as 'the best straight line through the origin'.

(c) In Figs 3.1 and 3.2 we now have two straight lines and a curve. Which of these three will give the best predictions of moisture content for belt speeds between 0 and 4 cm/s?

(d) Which of the three lines will give the best predictions of moisture content for belt speeds between 2.6 and 3.4cm/s?

While reflecting on Figs. 3.1 and 3.2 we need to keep in mind Dr Hubard's objectives. With this particular dryer he intends to use belt speeds around 3.0 cm/s. He certainly does not anticipate using belt speeds less than 2.0 or greater than 4.0 cm/s. Thus he wants an equation which accurately describes the relationship between belt speed and moisture content within this range. The obvious choice is the regression equation $y = -1.54 + 1.61x$. The curve is ruled out because he cannot obtain the equation of the curve from his calculator. The straight line through the origin is ruled out because it does not fit the data very well.

The fact that the regression line is the worst of the three at very low belt speeds is irrelevant to Dr Hubard's objectives. Within the range of the data the regression line is the best of the three and he does not intend to go far beyond the range of belt speeds in the data. I hope that Figs. 3.1 and 3.2 illustrate the great danger associated with extrapolation in this, or any other, situation.

3.4   HOW WELL DOES THE LINE FIT THE DATA?

I have described the line in Fig. 3.1 as the 'best' straight line. Later, we shall examine the criterion that is normally used when deciding whether one line is better than another. For the moment, I would like to introduce some simple methods of quantifying how well the line fits the data. It is wise to bear in mind that, even though a line is the 'best', it may not be good enough for any particular purpose.

Having typed the data into your calculator (I am assuming that you have now obtained one), you can obtain not only the regression equation but also the 'correlation coefficient'. This will tell you how well the line fits the data. The correlation coefficient can be negative or positive. It always has the same sign as the slope and will have a value between -1 and $+1$. Either of these extreme values indicates a perfect fit as we see in Figs. 3.3 and 3.4.

Midway between these two extremes we have a correlation coefficient of zero. This may indicate that there is no relationship between the two variables. Such is the case in Fig. 3.5, for example. However, a correlation coefficient of zero could arise

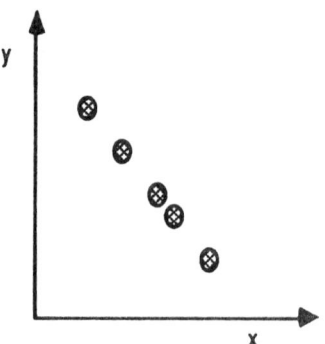

Fig. 3.3 — Correlation=−1.

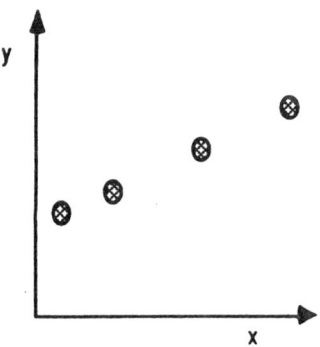

Fig. 3.4 — Correlation=1.

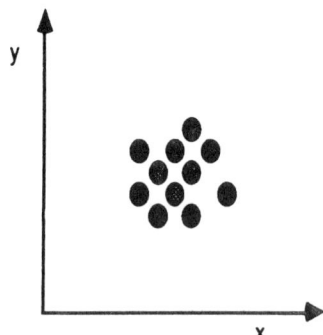

Fig. 3.5 — Zero correlation.

even if there was a relationship, as we can see in Fig. 3.6. This is a reminder that the correlation coefficient is a measure of linear relationship.

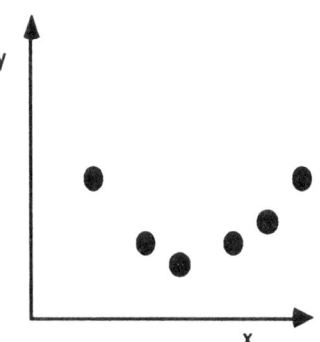

Fig. 3.6 — Zero correlation.

The data plotted in Fig. 3.1 have a correlation coefficient of 0.867 94. It is positive because the slope of the regression line is positive. The fact that it is so close to the maximum possible correlation is rather impressive. On reflection, however, Dr Hubard **should** expect a high positive correlation as he knows without doubt that increasing the belt speed will increase the moisture content. The purpose of the investigation was to quantify the relationship, not to prove that it exists. 'Why then', you may ask, 'is Dr Hubard fiddling around with the correlation coefficient?'

He now has the regression equation, which quantifies the relationship and allows him to predict the moisture content that he can expect with any belt speed setting. Why does he want to know how well the line fits the data? Perhaps he wants to know because he realizes that a line which fits better will give more precise predictions. In fact he needs the correlation coefficient if he is to take the next step in the analysis. This is to calculate confidence limits for the moisture content he can expect with a specified belt speed.

Before we calculate these confidence limits, let me introduce an alternative measure of goodness of fit, which may appeal to you more than the correlation coefficient. It is known as the 'percentage fit', and is very easily calculated from the formula

$$\text{percentage fit} = 100(\text{correlation})^2$$

Obviously, a correlation coefficient of −1 or +1 will give 100% fit, while zero correlation will give 0% fit. The data for belt speed and moisture content had a correlation coefficient of 0.867 94, which corresponds to a percentage fit of 75.3%. We say that 75.3% of the variation in moisture content is accounted for by the variation in belt speed. The remaining 24.7% of the variation in moisture content from batch to batch is due to other causes. Dr Hubard should reflect on what these other causes might be. Perhaps they include measurement error and/or changes in other variables which were not controlled.

Exercise 3.3

In this exercise I would like you to consider regression and correlation in a totally different context. Tom Reynolds is manager of the packaging department of Optico plc, which manufactures laboratory glassware of the highest quality. He is concerned that recent breakages during transit may have resulted from the carelessness of his packaging girls. He suspects that they sometimes cut corners in order to pack orders more quickly and thus to increase their bonus payments.

Table 3.2 contains the packaging times for nine consignments of spiral tubes packed by one particular operator. These data were recorded by Reynolds during an otherwise uneventful morning. He hopes to use the data to assess the times which need to be allowed for the packaging of consignments containing different numbers of tubes.

Table 3.2 — Packaging times for nine consignments

Consignment	A	B	C	D	E	F	G	H	I
Number of items	2	1	2	3	3	1	2	3	1
Time taken (min)	21	16	24	24	28	13	19	27	17

(a) Which of the two variables in Table 3.2 would you regard as the independent variable?
(b) Type the data into your calculator to obtain the slope and intercept of the regression equation.
(c) Plot a scatter diagram and draw on it the best straight line.
(d) Obtain the correlation coefficient from your calculator and calculate the percentage fit.
(e) Explain to Tom Reynolds what meaning can be attached to the slope, the intercept and the percentage fit.
(f) What time would you allow for this girl to pack a consignment of three items, if you wished to be very confident that she would be able to complete the task within the time allowed?

3.5 HOW ACCURATE ARE THE PREDICTIONS?

Earlier I pointed out that an equation which fitted the data well could be expected to give accurate predictions. Thus, with a correlation of 0.87, Peter Hubard can expect much better predictions of moisture content than he would have obtained if the correlation had been 0.1, say. Let us now calculate confidence limits for the moisture content that can be expected with any specified belt speed.

The width of confidence interval will depend on the correlation coefficient. Before we can obtain confidence limits we must first calculate what is known as the 'residual standard deviation'. Like all standard deviations it is a measure of variation.

It quantifies the scatter of the points about the regression line. If all the points lay on the line, the residual standard deviation would be equal to zero. It is calculated as follows:

$$\text{residual standard deviation} = (\text{SD of } y)\sqrt{[(1-r^2)(n-1)/(n-2)]}$$

where r is the correlation coefficient and n is the number of points. Note that the residual standard deviation has $n-2$ degrees of freedom.

For the belt speed and moisture data in Table 3.1 we have $r=0.8679$ and $n=10$ and the standard deviation of moisture content is 0.56502%. Substituting into the formula we obtain

$$\text{residual standard deviation} = (0.5602)\sqrt{[1-0.8679^2)(9)/(8)]}$$
$$=0.2952\%$$

Whenever we fit a regression line we are likely to find that the residual standard deviation is both useful and meaningful. In general terms the residual standard deviation is the standard deviation we could expect to obtain for the dependent variable if the independent variable were constant. Thus we could expect the standard deviation of moisture content to be about 0.295% if we produced many batches using the same belt speed. Note that the residual standard deviation is less than the standard deviation of moisture content (0.5602%) for the ten batches in Table 3.1. The residual standard deviation will always be less than the standard deviation of the dependent variable, unless the correlation is exactly zero. The relationship between the various measures of fit can be seen in Table 3.3.

Table 3.3 — Residual standard deviation is related to the correlation and the SD of y

Correlation coefficient	−1	0	+1
Percentage fit	100%	0%	100%
Residual standard deviation	Zero	SD of y	Zero

I will have more to say about the residual standard deviation later, when we examine residual variation in more detail. Let us now make use of the residual standard deviation (RSD) to calculate confidence limits. Whenever we fit a regression line there are several sets of confidence limits that might be of interest. The four formulae that we use are rather similar in some respects. I have listed them below so that you can see the similarity. I hope that the close juxtaposition of so many algebraic symbols will not disturb you.

Confidence limits for the true slope:

$$b \pm t(\text{RSD})\sqrt{(1/\text{SXX})} \tag{F1}$$

Confidence limits for the true intercept:

$$a \pm t(\text{RSD})\sqrt{[(1/n) + (\bar{x}^2/\text{SXX})]} \tag{F2}$$

Confidence limits for the mean of y corresponding to a specified value of x:

$$(a + bX) \pm t(\text{RSD})\sqrt{\{(1/m) + (1/n) + [(X - \bar{x})^2/\text{SXX}]\}} \tag{F3}$$

Confidence limits for the value of x corresponding to a mean value of y:

$$(Y - a)/b \pm [t(\text{RSD})/b]\sqrt{\{(1/m) + (1/n) + [(Y - \bar{y})^2/(b^2\text{SXX})]\}} \tag{F4}$$

In all the above formulae, n is the number of points to which the regression line was fitted, a is the intercept of the regression line, b is the slope of the regression line, t is taken from Table ST1 with $n-2$ degrees of freedom, RSD is the residual standard deviation, SXX is equal to (SD of $x)^2(n-1)$, X is a specified value of x, Y is the mean of m observations of y and m is the number of observations of y which will be used in the prediction of x (if m is equal to infinity Y will be a population mean).

The four formulae for confidence limits are labelled (F1)–(F4) for ease of reference. You will have noticed that the formulae become more complex as you read down the list. Let us, therefore, start with the simplest formula, which gives confidence limits for the true slope. 'What exactly is the true slope?', you may ask.

We calculated the slope of the best straight line to be 1.61, and then we suggested to Dr Hubard that the moisture content would increase by 1.61% if he increased the belt speed by 1 cm/s. However, it would be very unwise to carve this number on a tablet of stone, for it is only an **estimate** of the true effect. If Dr Hubard added to his data the results from an additional batch, our estimate would probably change. Furthermore, if he produced another ten batches the new data would almost certainly give a different estimate. The 'true slope' is the estimate we would obtain if we examined an unlimited number of batches.

If we substitute, into formula (F1), $n=10$, $b=1.61$, $t=2.31$ (with 8 degrees of freedom), RSD$=0.2952$ and SXX$=0.8240$, we obtain a 95% confidence interval of 1.61 ± 0.75. Thus we can be 95% confident that the true slope is between 0.86 and 2.36. In practical terms this means that an increase in the belt speed of 1 cm/s might increase the moisture content by as much as 2.36% or by as little as 0.86%.

Using formula (F2) we can obtain 95% confidence limits for the true intercept. If you work through the calculation you should obtain -1.54 ± 2.29. We notice that this confidence interval includes zero, which suggests that the true regression line might pass through the origin. However, as Dr Hubard showed no interest in the calculated intercept, he is unlikely to be interested in the confidence interval either, so we shall proceed to the next formula. Before we do so, however, I should point out that the confidence interval for the true intercept would have been much narrower if the belt speeds in Table 3.1 had been closer to zero.

Exercise 3.4

In this exercise we shall give further consideration to the data in Table 3.2 which were used in Exercise 3.3.

(a) Calculate the residual standard deviation. (If you have cleared your calculator since doing Exercise 3.3, you will need to type in the data again.)
(b) Calculate 95% confidence limits for the true slope.
(c) Calculate 95% confidence limits for the true intercept.
(d) Explain to Tom Reynolds in practical terms what meaning can be attached to the residual standard deviation and the confidence limits. (Tom Reynolds is the manager of the packaging department who collected the data.)

Formula (F3) is the one which really interests Dr Hubard. This gives confidence limits for the mean of y for a specified value of x. When using formula (F3) we would set m equal to **infinity** if we wished to predict the long-term mean, or population mean. Alternatively we would set m equal to unity if we are interested in one occasion only; the next batch, say. Suppose, for example, Dr Hubard intends to run the dryer with the belt speed set at 3.0 cm/s. Substitution of $x=3.0$ into the regression equation gives a predicted moisture content of 3.3%. Thus we could claim that the mean moisture content of future batches will be 3.3% if the belt speed is set at 3.0 cm/s. Obviously this prediction is likely to be in error. Thus it would be helpful to have confidence limits which indicate just how low or how high the mean moisture content might be. These limits can be obtained from formula (F3).

Substituting $a=-1.43355$, $b=1.606796$, $t=2.31$, RSD$=0.2952$, $n=10$, $X=3.0$, $\bar{x}=3.04$, SXX$=0.8240$ and putting m equal to infinity, we obtain 95% confidence limits as follows:

$$(a+bX)\pm t(\mathrm{RSD})\sqrt{\{(1/m)+(1/n)+[(X-\bar{x})^2/\mathrm{SXX}]\}}$$

$$=3.3\pm 2.31\,(0.2952)\sqrt{\{0+(1/10)+[(3.0-3.04)^2/0.8240]\}}$$

$$=3.3\pm 0.6819\sqrt{(0+0.1+0.00194)}$$

$$=3.3\pm 0.2$$

Thus we can be 95% confident that the mean moisture content of future batches will be between 3.1% and 3.5% if a belt speed of 3.0 cm/s is used.

An important point to note about formula (F3) is that the width of the confidence interval depends on the value of X. With X close to $\bar{x}$ we obtain a narrow interval, but as X deviates from the centre of the data the interval widens. If we substitute a value for X which is well outside the belt speeds used in the trial, the interval will be very wide. This point is illustrated by the confidence limits in Table 3.4. These are plotted to give the confidence bands in Fig. 3.7.

Table 3.4 — Confidence intervals differ in width

Belt speed x	Predicted moisture content $a+bx$	Confidence limits for the true mean moisture content
5.0	6.49	5.00 to 7.98
4.0	4.88	4.13 to 5.63
3.04	3.34	3.12 to 3.56
2.0	1.67	0.86 to 2.48
1.0	0.06	−1.49 to 1.61
0.0	−1.54	−3.83 to 0.75

In our use of formula (F3) we have set m equal to infinity because we were interested in the **long-term** consequences of using a specified belt speed. Suppose Dr Hubard is particularly interested in one particular batch, perhaps the next batch. To obtain confidence limits for the moisture content of one batch we repeat the calculation using m equal to unity. As individual batches are scattered about the true mean, we should not be surprised to find that putting m equal to unity gives a wider interval than we would obtain with m equal to infinity. For a belt speed of 3.0 cm/s with $m=1$ we find 95% confidence limits of 2.6% and 4.0%. Compare this with the confidence limits of 3.1% and 3.5% we obtained earlier with m equal to infinity. Thus, if we use a belt speed of 3.0 cm/s, we can be 95% confident that the long-term mean moisture content will be between 3.1% and 3.5%, but the moisture content of any particular batch might be as low as 2.6% or as high as 4.0%.

The last of the four formulae can be used to obtain confidence limits for the true value of x corresponding to a mean value of y. As with formula (F3) we put m equal to **infinity** if we are using the population mean of y (i.e. the long-term mean), but we put m equal to unity if we are basing our prediction of x on a single observation of y. Perhaps these confidence limits are not what Dr Hubard really wants. He would like an answer to the question, 'What is the highest belt speed that I can use and still be confident of getting no more than 3.5% moisture?' The simplest way to obtain an answer is to calculate 90% confidence limits using formula (F4), and then to quote only the lower limit. Pursuing this strategy we conclude that a belt speed of 3.03 cm/s will give a 95% chance of having no more than 3.5% moisture in any batch.

By now you have probably seen more than enough of the confidence interval formulae. Before we leave them, however, you may be interested to know which of the four are most widely used. Certainly, formula (F1) is more useful than (F2) because there are many situations in which the researcher is interested in the slope but not the intercept. Of the other two, formula (F3) is much more frequently used than (F4). Furthermore, both formulae are more likely to be used with m equal to infinity, rather than with m equal to unity. There is, however, one notable exception. Formula (F4) is important in analytical chemistry when predicting the concentration of an unknown sample using a calibration line. In this application of regression analysis, m would be equal to the number of determinations made on the unknown sample.

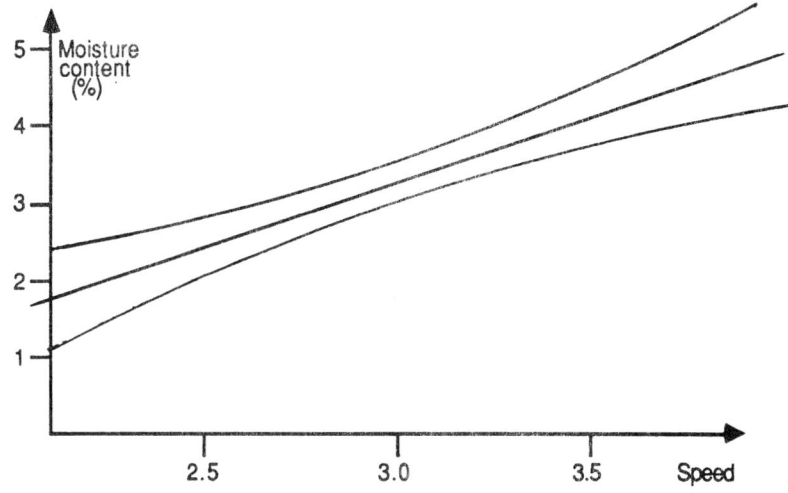

Fig. 3.7 — Confidence bands for the true mean moisture.

Exercise 3.5

In this exercise we shall refer once more to the data in Table 3.2 which were first used in Exercise 3.3.

(a) Calculate 95% confidence limits for the average packing time for consignments containing three items.
(b) Calculate 95% confidence limits for the time needed to pack the next consignment which contains three items.
(c) How would you use formula (F4) to help Tom Reynolds acquire a better understanding of the relationship between packing time and size of consignment?

3.6 ASSUMPTIONS AND RESIDUALS

You will realize, having studied Chapters 1 and 2, that all statistical techniques are based on assumptions. You will also realize that it is dangerous to use any technique without being aware of these assumptions. Furthermore, it is desirable that you should be conversant with the methods that can be used to check whether or not the assumptions are violated. Let us now examine the assumptions underlying simple linear regression analysis. As we do so you will probably acquire a better understanding of what we have discussed so far. The assumptions are:

(a) The true relationship between the two variables is linear.
(b) There is no error in the values of the independent variable.
(c) The errors in the values of the dependent variable are a random sample from a population of errors.

(d) The population of errors has a normal distribution, with a constant standard deviation.

The first assumption implies that we could be misled by the best straight line, or the regression equation, if the true relationship were a curve. This is a rather obvious point. However, we should not forget that any curve can be closely approximated by a straight line, if we are interested in only a **small section** of the curve. The analytical chemist knows full well that a calibration curve is virtually straight if he uses only a small range of concentrations.

Assumptions (b) and (c) taken together tell us that there are errors in y but not in x. With the Vulcatuf data in Table 3.1 this assumption would appear to be satisfied. The belt speed can be set very accurately and maintained at any desired value. It is, therefore, reasonable to regard the x values as error free. The moisture content, in contrast, is determined in the analytical laboratory to which samples are sent. Thus the y values are subject to sampling and testing errors. Peter Hubard knows that these errors are not negligible. Furthermore, there may be variation in moisture content due to factors beyond our control. This additional variation in y will also be classed as error, because it is not due to changes in the independent variable, belt speed.

Assumptions (b) and (c) are very important if we wish to estimate the parameters of a functional relationship between the two variables. However, they are not so important if we simply wish to use the fitted equation to make predictions.

The final assumption tells us that the regression analysis will only be valid if the population of errors has a normal distribution. I mentioned the normal distribution in Chapter 1 and we shall discuss it again, in detail, in a later chapter. However, your knowledge or ignorance of normal distributions does not create the major difficulty in the discussion of this assumption. Far more important is our inability to separate the error from the true value. This is an important point which requires careful consideration.

In any regression analysis, each value of y can be regarded as the sum of two parts. Furthermore, the split into two parts can be achieved in two ways:

(a) Observed value of y=true value of y+error: the true value of y is equal to $A+Bx$.
(b) Observed value of y=predicted value by y+residual: the predicted value of y is equal to $a+bx$.

Every one of the ten y values in Table 3.1, or the nine y values in Table 3.2, could be broken down into two parts. Thus we could separate the error in each observed y value if only we knew the true value. A true value could be calculated if we knew the true intercept (A) and the true slope (B), but of course we do not. However, the calculated intercept (a) is an estimate of the true intercept (A) and the calculated slope (b) is an estimate of the true slope (B). Thus, to estimate the 'true value of y', we shall use the 'predicted value of y', and to estimate the errors we will use the residuals.

It might be a good idea to read the two previous paragraphs again, very carefully. Then you might care to inspect Table 3.5 which contains residuals for the data in Table 3.1.

Table 3.5 — Residual=observed y−predicted y

Belt speed x	Observed moisture content y	Predicted moisture content $a+bx$	Residual $y-(a+bx)$
2.9	2.8	3.12	−0.32
3.2	3.3	3.60	−0.30
2.6	2.3	2.63	−0.33
3.4	3.7	3.92	−0.22
3.1	3.4	3.44	−0.04
3.1	3.7	3.44	0.26
3.4	4.1	3.92	0.18
2.6	2.8	3.63	0.17
3.3	3.9	3.76	0.14
2.8	3.4	2.95	0.45
Mean	3.34	3.34	0.00

To calculate a residual we subtract the predicted moisture content from the observed moisture content. (A predicted y value can be obtained very easily on some calculators by typing in the x value and then pressing the $\hat{y}$ key). You will notice that some residuals are negative while others are positive and the sum of the residuals is zero. A negative residual tells us that the observed moisture content of the batch is less than we would expect with that belt speed. The residuals could also be obtained from the scatter diagram by measuring the distance of each point from the fitted line.

In Fig. 3.8 the residuals are the lengths of the vertical lines which join the

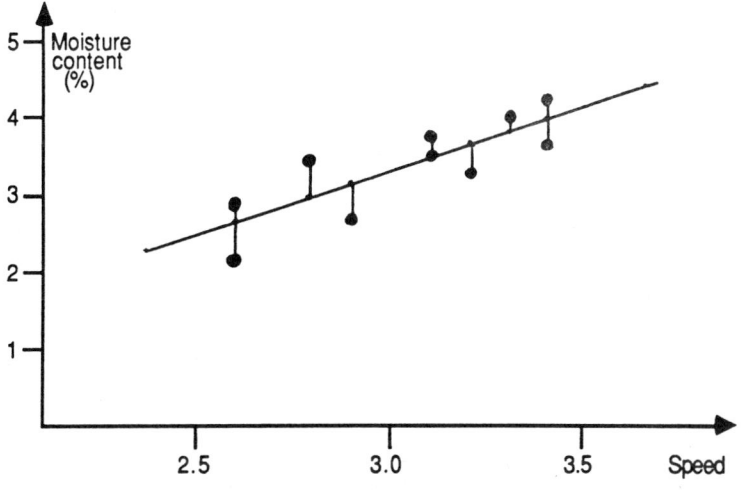

Fig. 3.8 — Residuals and predicted values.

observed points to the predicted points. The five points below the regression line represent the batches which have negative residuals. If the true relationship is linear, each residual in Table 3.5 will be an estimate of the error in the observed moisture content for that batch. By examining the residuals we can check the assumptions. Fig. 3.9 is a dot plot of the residuals.

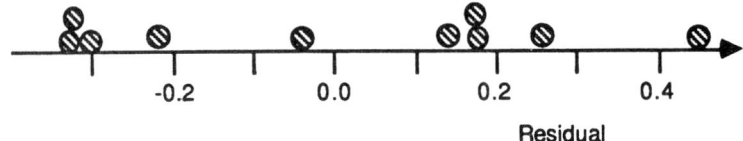

Fig. 3.9 — Distribution of the residuals.

If **all** of the assumptions are satisfied the residuals should resemble a random sample from a normal distribution. Thus, if we find any pattern in the residuals, we must suspect that one or more of the assumptions is violated. An inspection of Fig. 3.9 does not convince me that I am looking at a sample from a normal distribution. Instead of the dots' being crowded around the mean with one or two at the extremes, we find that the dots fall into two clusters.

There is further evidence of an abnormal pattern if we examine the residuals in Table 3.5. We see that the first five residuals are negative and the last five are positive. As the batches are listed in the order of manufacture, this implies that the earlier batches have lower moisture content than the later batches, even after we have taken account of any difference in belt speed. Perhaps some factor, other than the belt speed, changed after the fourth or fifth batch. A plot of the residuals against their batch numbers may throw light on the problem (Fig. 3.10).

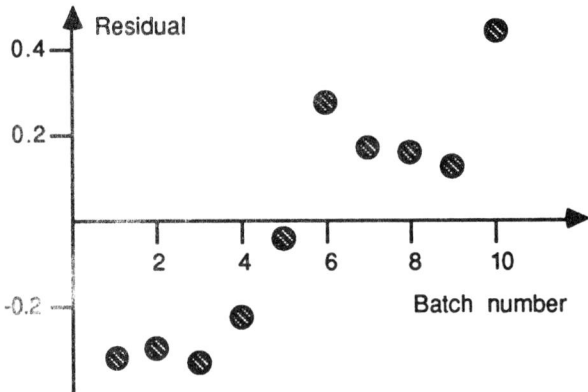

Fig. 3.10 — Residuals plotted against batch number.

Plotting graphs of the residuals is very helpful when we are checking the regression assumptions. It is common practice to plot the residuals against the independent variable, against the dependent variable and/or against the observation number. I am of the opinion, however, that all of these graphs are improved if we add to each of the residuals the mean of the dependent variable. This simple operation converts the residuals into what could be described as adjusted values. Adding the mean moisture content, 3.34%, to each residual gives us an adjusted moisture content for each batch (Fig. 3.11). These adjusted values are the moisture content we

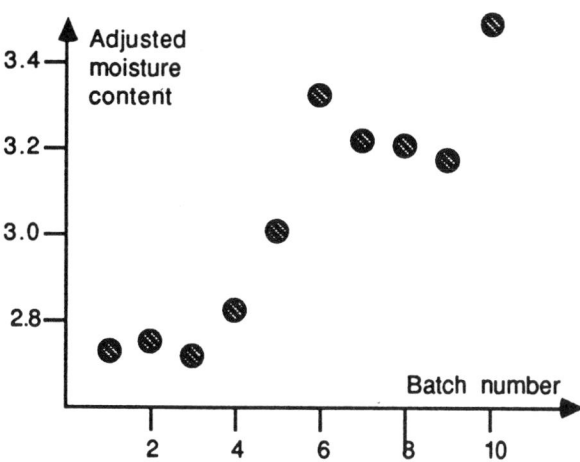

Fig. 3.11 — Adjusted moisture plotted against batch number.

could reasonably have expected if all ten batches had been dried at the average belt speed, 3.04 cm/s. When calculating these adjusted moisture values we in effect take the actual moisture content and increase it or decrease it to make allowance for the different belt speeds used.

Whether we examine Fig. 3.10 or the more meaningful Fig. 3.11, we receive the same impression — that the moisture content increased after the fourth or fifth batch, and that this increase was not due to a change in belt speed. In the light of this discovery it is no longer reasonable to assume that a linear relationship prevailed throughout the experiment. It would be interesting, therefore, to fit **two** straight lines, one for the earlier batches and one for the later batches.

Exercise 3.6

In this exercise we shall make use of the data for belt speed and moisture content in Table 3.1.

(a) Using the data for batches 627 to 631 inclusive, obtain the intercept, slope and correlation coefficient. Draw the regression line on Fig. 3.8.
(b) Using the data for batches 632 and 636 inclusive, obtain the intercept, slope and correlation coefficient. Draw the regression line on Fig. 3.8.

(c) By measuring the distance of each point from the appropriate line, estimate each of the ten residuals.
(d) Draw a dot plot of the ten residuals. Do they resemble a sample from a normal distribution?
(e) Record any reservations you may have about what you have been asked to do in parts (c) and (d). (Hint: compare the size of the residuals with the size of the rounding errors in the moisture content values in Table 3.1.)

3.7   CORRELATION CAN BE MISLEADING

For any set of data the correlation coefficient must have a value between -1 and $+1$. Whenever we find a correlation close to either extreme we may be tempted to conclude that there is a cause and effect relationship between the two variables. This temptation **should** be resisted until:

(a) a scatter diagram has been examined, and
(b) thought has been given to how the cause–effect relationship might operate.

In Fig. 3.12 we can see how the correlation coefficient can be strongly influenced

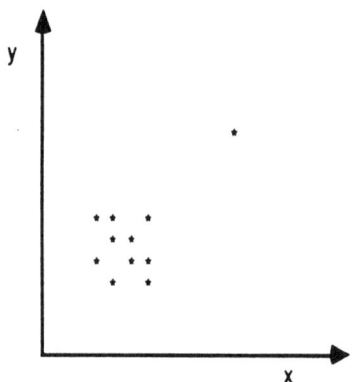

Fig. 3.12 — Correlation=0.9.

by one observation. In Fig. 3.13 we can see that within each group of points there is zero correlation, but the two groups taken together give a very high correlation. In Fig. 3.14 we can see that there would be a very high correlation, but for the one point which does not seem to fit the pattern displayed by the others. Let these three diagrams serve as a reminder that it is dangerous to interpret a correlation coefficient, or a regression analysis, without examining a scatter diagram.

Fig. 3.15 conveys a very different message. It tells us that a set of data containing

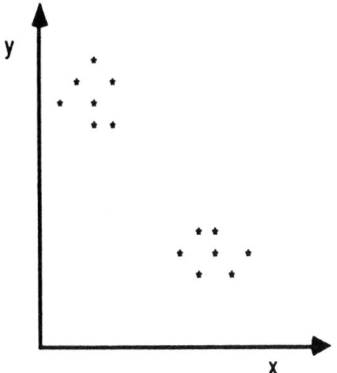

Fig. 3.13 — Correlation = −0.9.

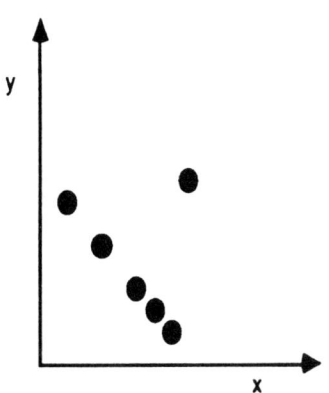

Fig. 3.14 — Correlation=0.

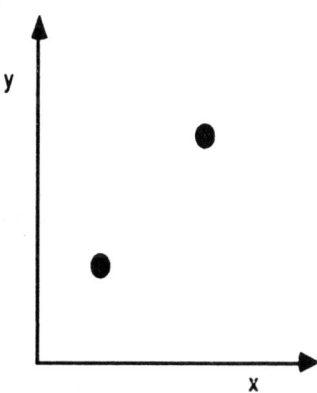

Fig. 3.15 — Correlation=1.

only two points **could** give a correlation coefficient of $+1$. This is rather obvious. However, the diagram will have served its purpose if it helps you to realize the two observations will almost certainly give a correlation of $+1$ or -1, **even if there is no relationship between the two variables**.

Perhaps you have no intention of basing decisions on only two points. Consider, then, what would be the result of adding a third point at random to Fig. 3.15. Would the correlation fall to zero? It seems very unlikely. If you think about the scatter diagrams with very few points I am sure you will realize that small data sets are quite likely to give high correlation coefficients even if there is no true relationship between the variables. Because of this possibility, it is wise to check a calculated correlation against the appropriate value from Table ST3 before you present your conclusions to a suspecting audience. (Table ST3 is one of the statistical tables at the end of the book.)

We can see in Table ST3 that, with only six points on the scatter diagram, we require a calculated correlation greater than 0.811 to prove, beyond reasonable doubt, that a relationship exists. With a much larger sample of 60 points, we need to exceed only 0.254 in order to conclude with 95% confidence that the variables are related. The required values in Table ST3 obviously become smaller as the sample size increases. This is because, as we noted earlier, a smaller sample is more likely to give a spurious correlation.

Peter Hubard passed ten batches of Vulcatuf through the dryer to obtain the data in Table 3.1. The correlation between belt speed and moisture content was 0.868. For a sample size of ten the required value from Table ST3 is 0.632. Thus we can reasonably conclude that there is a relationship between belt speed and moisture content. Of course, Peter Hubard is unlikely to be very impressed by this conclusion, as he already knew that this relationship existed before he did the experiment. The purpose of the investigation was to obtain the regression equation.

Before we leave this topic let me give you a further caution about the interpretation of correlation coefficients. The finding of a correlation which is greater than the required value from Table ST3 does not prove that there is a cause–effect relationship between the two variables. In fact correlation **alone** does not tell us anything about cause and effect. A simple example will clarify this point.

If you were to tabulate the annual sales of whisky in Great Britain for each year from 1970 to 1980 alongside the average salary of Methodist ministers for each of these years, you would find that these data gave a very high and positive correlation. The correlation coefficient would certainly be greater than the required value from Table ST3. Would you then conclude that whisky sales could be decreased by reducing ministers' salaries? Would you also conclude that Methodist ministers could be made more affluent by the British people's buying more whisky? Of course not.

3.8   A SUMMARY OF THE IMPORTANT POINTS

(1) Exploring the relationship between two variables is a common activity in the chemical industry, as we strive to gain a better understanding of complex processes, with a view to achieving better performance.

(2) Plotting a scatter diagram and fitting the best straight line, or regression line, are essential steps in establishing a relationship.

(3) The regression equation and the correlation coefficient are easily obtained from a suitable calculator.

(4) In many situations we are very interested in the slope of the regression line but we have little or no interest in the intercept. The slope tells us how much the dependent variable will change if we increase the independent variable by one unit.

(5) The correlation coefficient and the percentage fit tell us how well the line fits the points.

(6) Even if the true relationship is a curve, the regression line may fit very well, within the range of the data. However, extrapolation beyond the data can produce predictions which are very inaccurate, even if the true relationship is linear).

(7) The residual standard deviation (RSD) quantifies the variability in y for a fixed value of x. It is most easily calculated using the correlation coefficient (r):

$$RSD = (SD \text{ of } y)\sqrt{[1-r^2)} (n-1)/(n-2)]$$

(8) The calculated slope (b) is an estimate of the true slope (B). Confidence limits for the true slope are given by:

$$b \pm t\,(RSD)\sqrt{(1/SXX)}$$

(9) Having fitted the regression equation we might wish to predict the value of the dependent variable that will result from a particular value of the independent variable. Confidence limits for the mean of y corresponding to a specified value of x are given by:

$$(a+bX) \pm t(RSD)\sqrt{\{(1/m)+(1/n)+[(X-\bar{x})^2/SXX]\}}$$

If we wish to predict the long-term mean, i.e. the population mean, we put m equal to infinity. If we wish to predict a single occurrence we put m equal to 1.

(10) Whenever we fit a regression equation we should consider the assumptions underlying this technique. If all the assumptions are satisfied the residuals will resemble a random sample from a normal distribution. Any pattern in the residuals is an indication that one or more of the assumptions may be violated.

(11) When choosing which of the two variables we shall regard as the independent variable, we need to consider:
   (a) whether there is a cause–effect relationship;
   (b) which variable we wish to predict;
   (c) which variable has the larger errors.

(12) Before we assert that a relationship exists between two variables, we should compare the calculated correlation coefficient with the required value from Table ST3. However, even if the correlation exceeds the required value, this **alone** does not prove that a cause–effect relationship exists.

3.9   ADDITIONAL EXERCISES

Exercise 3.7

We have said very little about the throughput data in Table 3.1, although they are just as important as the data on moisture content. Peter Hubard wishes to quantify the relationship between belt speed and throughput, so that he can estimate the cost of attempting to achieve any particular level of moisture content.

(a) Using the data from Table 3.1 find the slope and intercept of the regression line. (I assume that you will use a calculator or a computer to do this.)
(b) Obtain the correlation coefficient and calculate the percentage fit.
(c) Using the $\hat{y}$ key in your calculator, or by some other means, obtain a predicted throughput for each of the ten batches.
(d) Calculate a residual for each batch by subtracting the predicted throughput from the actual throughput. Do not ignore the negative signs.
(e) Plot the residuals against the belt speed. Do you see any pattern in this plot?
(f) What conclusions do you draw about the usefulness of the regression equation?

Exercise 3.8

Chris Mawgan works in the Research and Development Department of Colour Chemicals Ltd. In earlier trials he has shown that the addition of triazone during the manufacture of digozo blue pigment reduces the level of impurity in the final product. He has just finished an additional trial designed to quantify more precisely the effect of adding triazone (Table 3.6).

Table 3.6 — Seven batches of digozo blue

Batch number	421	422	423	424	425	426	427
Triazone (kg)	1.0	0.5	2.5	3.5	2.0	1.5	3.0
Impurity	6.7	6.4	3.2	2.3	4.7	5.0	4.1

(a) Which variable would you select to be the independent variable?
(b) Calculate the slope and intercept of the regression line.
(c) Calculate the correlation coefficient and the percentage fit.
(d) Calculate the residual standard deviation.
(e) Explain to Dr Mawgan what meaning can be attached to the slope, the intercept and the residual standard deviation.
(f) Calculate 95% confidence limits for the true slope.
(g) Calculate 95% confidence limits for the long-term mean impurity that could be expected if 3.0 kg of triazone are added to all future batches.
(h) Dr Mawgan is not familiar with confidence limits. Explain to him the usefulness of what you calculated in parts (f) and (g).

3.10 WORKED SOLUTIONS

Solution to Exercise 3.1

(a) In Table 3.1 we can see that no batches had a moisture content of 3.5%. However, two batches had moisture content of 3.4%, when belt speeds of 3.1 and 2.8 cm/s were used. Furthermore, two batches had a moisture content of 3.7%, when belt speeds of 3.1 and 3.4 cm/s were used. Assessing the data from these four batches might lead you to suggest that a belt speed of approximately 3.1 cm/s would give a moisture content of 3.5%. This assessment is based on only four of the ten batches. Perhaps it would be wiser to take account of all 10 batches, which can be achieved by plotting a scatter diagram.

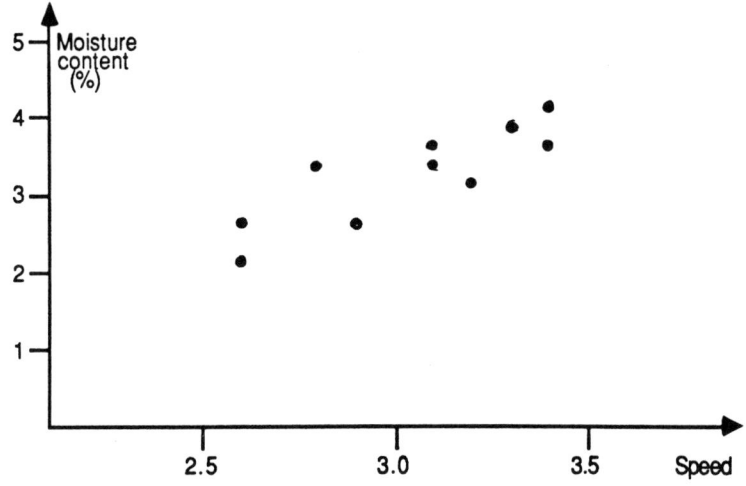

Fig. 3.16 — Relationship between belt speed and moisture content.

   By drawing on Fig. 3.16 a straight line which fits the data as closely as possible, we could use this line to predict the belt speed needed to achieve any specified level of moisture content. If you draw on the scatter diagram what you consider to be the 'best straight line' you might well find a predicted belt speed of about 3.1 cm/s.

(b) While looking at the scatter diagram we cannot fail to notice the considerable scatter of the data points around the line. Thus we realize that two batches dried at the same belt speed may not have exactly the same moisture content. Hence, we would need a lower belt speed than that predicted in part (a) if we wanted 95% of batches to have a moisture content of 3.5% or less. If you drew onto the scatter diagram a line which was parallel to the best straight line but a little higher so that all (or 95%) of the data were below this line, you could use this to predict the required moisture content. Perhaps your prediction would be about 2.8 or 2.9 cm/s.

(c) To estimate the throughput that would result from the use of any particular belt speed we need a second scatter diagram. Fig. 3.17 indicates that a belt speed of 3.1 cm/s would give a throughput of approximately 170 kg/h.

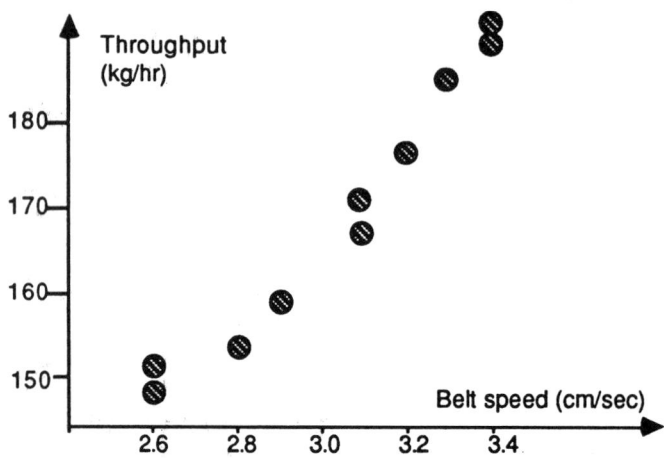

Fig. 3.17 — Relationship between throughput and belt speed.

Solution to Exercise 3.2
(a), (b)

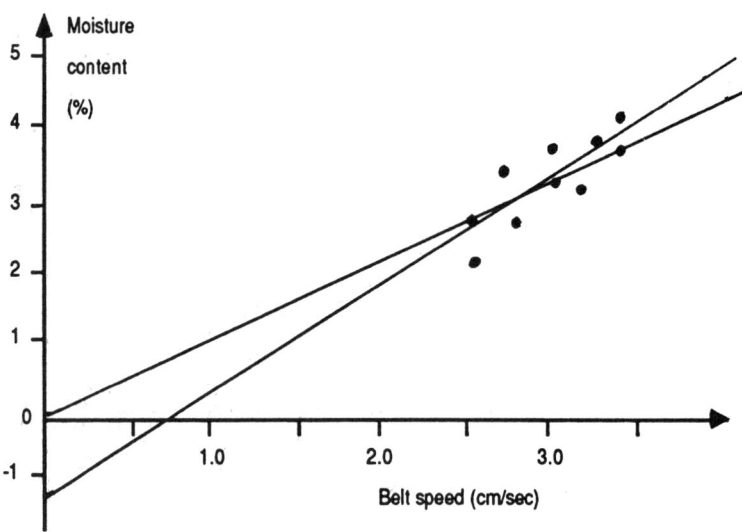

Fig. 3.18 — Best line and best line through the origin.

(c) At low belt speeds the worst predictions will come from the best straight line, $y=-1.54+1.62x$. Perhaps the curve gives better predictions than would the line through the origin.

(d) For belt speeds between 2.6 and 32.4 cm/s the worst predictions come from the line through the origin. There is little difference between the curve and the best straight line over this range of belt speeds.

Solution to Exercise 3.3
(a) The independent variable is the number of items. Obviously the time taken will depend on the number of items and not vice versa.
(b)   Intercept=10.0 min
                       Slope=5.50 min per item

The regression equation is $y=10.0+5.5x$, where y is the time taken (min) and x is the number of items.
(c)

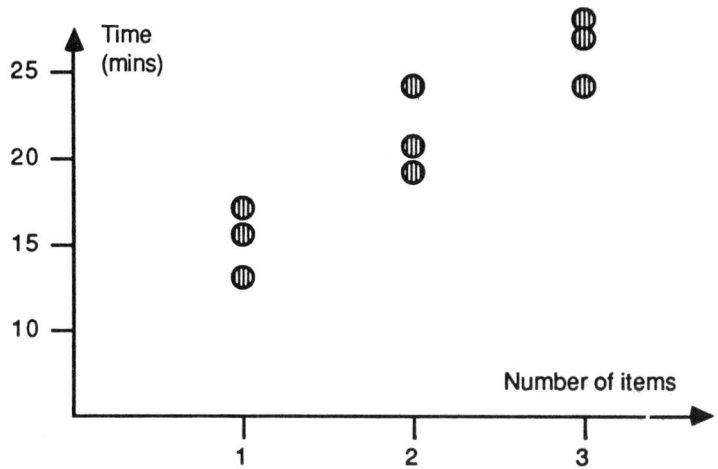

Fig. 3.19 — Relationship between time taken and number of items.

(d)   Correlation=0.925 27
                       Percentage fit=$100(0.925\,27)^2=85.6\%$
(e) The slope of the best straight line tells us how much extra time needs to be allowed for each extra item in the consignment. The intercept tells us how long it would take to pack a consignment which contained zero items. Clearly there are no such consignments. However, we can regard the intercept as the fixed part of the packing time and the slope as the variable part. Thus the time taken on average includes a fixed part of 10 min plus a variable part of 5.5 min for each item. The percentage fit tells us that 85.6% of the variation in packing time from consignment to consignment can be accounted for by the variation in the number of items. The other 14.4% is due to other factors such as fatigue, errors and faulty packets.
(f) Substitution of $x=3$ into the regression equation gives a predicted time of 26.5 min. However, if we allow only 26.5 min for a consignment of three items, there is a 50% chance that the girl will not have finished the task within the time

allowed. Thus we need to increase the time allowance. Perhaps an additional 3.5 min would be sufficient, giving a time of 30 min. We shall return to this point later.

Solution to Exercise 3.4

(a)   SD of $y=5.1478$ correlation$=0.92527$ $n=9$

residual standard deviation$=$(SD of y)$\sqrt{[(1-r^2)(n-1)/(n-2)]}$
$$=5.1478\sqrt{[(1-0.92527^2)(8)/7]}$$
$$=2.087 \text{ min.}$$

(b) $b=5.5$ SD of $x=0.86603$ SXX$=6.0$ $t=2.36$

95% confidence limits for the true slope are
$$b\pm t(\text{RSD})\sqrt{(1/\text{SXX})}$$
$$=5.5\pm 2.36\ (2.087)\sqrt{(1/6.0)}$$
$$=5.5\pm 2.01 \text{ min per item.}$$

(c) $a=10.0$ $\bar{x}=2.0$

95% confidence limits for the true intercept are
$$a\pm t(\text{RSD})\sqrt{[(1/n)+(\bar{x}/\text{SXX})]}$$
$$=10.0\pm 2.36\ (2.087)\sqrt{[(1/10)+(2.0^2/6.0)]}$$
$$=10.0\pm 4.31 \text{ min.}$$

(d) The residual standard deviation measures the variability in packing times for consignments containing the same number of items. It is an estimate of the standard deviation of packing times that we would find if this girl packed an infinite number of equal-size consignments. The confidence limits for the true slope tell us that adding an extra item to the consignment could increase its packing time by as little as 3.5 min or as much as 7.5 min. The confidence limits for the true intercept tell us that the 'fixed part' of the packing time is between 5.7 min and 14.3 min.

Solution to Exercise 3.5

(a) $a=10.0$ $b=5.5$ $t=2.36$ RSD$=2.087$ $n=9$
           m is equal to infinity $X=3$ $\bar{x}=2.0$ SXX$=6.0$

95% confidence limits for the mean packing time are
$$(a+bX)\pm t(\text{RSD})\sqrt{\{(1/m)+(1/n)+[(X-\bar{x})^2/\text{SXX}]\}}$$
$$=26.5\pm 2.36(2.087)\sqrt{\{0+(1/9)+[(3-2.0)^2/6.0]\}}$$
$$=26.5\pm 2.6 \text{ min}$$

Thus we can be 95% confident that the mean packing time for consignments of three items will be between 23.9 and 29.1 min.

(b) We repeat the calculation carried out in part (a) using m equal to 1, rather than infinity. 95% confidence limits for the packing time of the next consignment are:
$$(a+bX)\pm t(\text{RSD})\sqrt{\{(1/m)+(1/n)+[(X-\bar{x})^2/\text{SXX}]\}}$$
$$=26.5\pm 2.36(2.087)\sqrt{\{1+(1/9)+[(3-2.0)^2/6.0]\}}$$
$$=26.5\pm 5.6 \text{ min}$$

Thus we can be 95% confident that the packing time for the next consignment of three items will be between 20.9 and 32.1 min. Note that this confidence interval is much wider than that obtained in part (a), as we would expect.

(c) In my opinion, formula (F4) is not helpful in this situation. It is hard to imagine Tom Reynolds wanting to predict the number of items in a consignment from the packing time. It would be unwise to assume that all the formulae for confidence intervals will be useful in every situation.

Solution to Exercise 3.6

(a) Calculated intercept$=-2.29$
Calculated slope $=$ 1.774

The equation of the regression line is $y=-2.29+1.774x$
Correlation coefficient$=0.9797$

(b) Calculated intercept$=-0.89$
Calculated slope $=$ 1.469

The equation of the regression line is $y =-0.89+1.469x$.

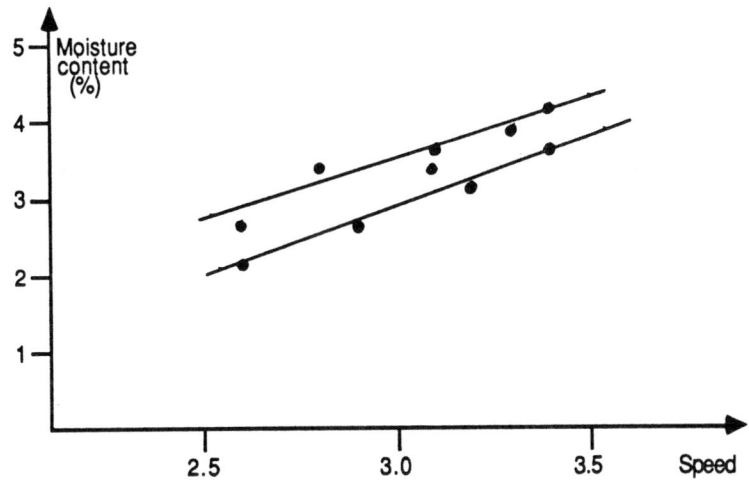

Fig. 3.20 — Separate lines for early and late batches.

(c)

Table 3.7 — Residuals from the two lines

Batch number	627	628	629	630	631	632	633	634	635	636
Residual	−0.05	−0.08	−0.02	−0.04	0.19	0.03	−0.01	−0.13	−0.06	0.17

(d)

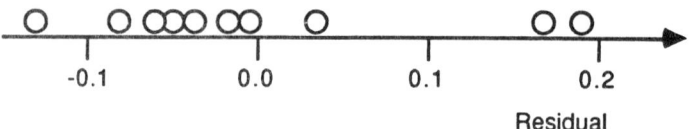

Fig. 3.21 — Residuals from the two lines.

These residuals would look very 'normal' but for the two large positive values. See part (e) for further comment.

(e) The moisture content determinations have been rounded to one decimal place. Thus the error in any one could be as large as ±0.05 simply because of rounding. Perhaps some errors are even greater. We see that some of the residuals in part (d) are less than 0.05 in magnitude. It is sometimes possible to fit more and more complex equations until the residuals become ridiculously small. This is known as 'overfitting'. Perhaps that point has now been reached.

Solution to Exercise 3.7
(a) Calculated slope   =53.9077
    Calculated intercept=5.8204
    The equation of the regression line is $y=5.82+53.908x$
(b) Calculated correlation coefficient=0.9775
    Percentage fit=100(correlation)^2
    $=100(0.9775)^2$
    $=95.6\%$
(c), (d)

Table 3.8 — Answers for Exercise 3.7(c), (d)

Belt speed x	Actual throughput y	Predicted throughput $a+bx$	Residual
2.9	159	162	−3
3.2	177	178	−1
2.6	151	146	5
3.4	193	189	4
3.1	167	173	−6
3.1	171	173	−2
3.4	191	189	2
2.6	148	146	2
3.3	186	184	2
2.8	154	157	−3

(e)

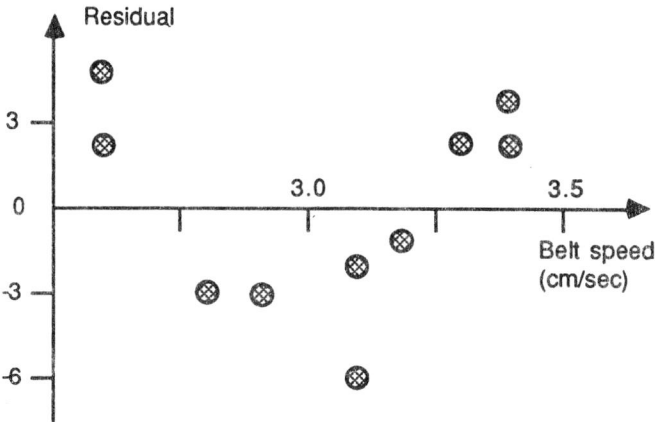

Fig. 3.22 — Plot of residuals against belt speed.

The points in Fig. 3.22 could be fitted well by a curve. Clearly there is a relationship between the residuals and the belt speed. Low belt speeds and high belt speeds have given positive residuals whilst medium belt speeds have given negative residuals.

(f) The pattern we observed in part (e) implies that the data could have been better fitted by a curve than by a straight line. If you draw your regression line on Fig. 3.17 you will see very clearly that a curve would fit much better even though the straight line has a percentage fit of 95.6%.

Solution to Exercise 3.8

(a) Triazone should be selected as the independent variable (x), with impurity as the dependent variable (y).

(b) Calculated slope $=-1.37857$
Calculated intercept$=7.38571$

The regression equation is, therefore, $y=7.39-1.379x$

(c) The calculated correlation coefficient is -0.9320.

(d) Residual standard deviation
$$=(\text{SD of } y)\sqrt{[(1-r^2)(n-1)/(n-2)]}$$
$$=(1.5976)\sqrt{[(1-0.9320^2)(6)/(5)]}$$
$$=0.6342\%$$

(Remember that the residual standard deviation is in the same units as the dependent variable.)

(e) The slope tells us that the increase of 1 kg in the weight of triazone added will give a reduction in impurity of 1.379%. The intercept tells us that we can expect 7.39% impurity if we do not add any triazone. The residual standard deviation of 0.6342% is much less than the standard deviation of y, 1.5916%, because we have such a high correlation coefficient. If a succession of batches were made using the **same** weight of triazone, we would expect the batch-to-batch variation in impurity to give a standard deviation of approximately 0.63%.

(f) Confidence limits for the true slope are given by:

$$b \pm t(\text{RSD})\sqrt{(1/\text{SXX})}$$
$$= 1.38 \pm 2.57\ (0.6342)\sqrt{(1/7.0)}$$
$$= 1.38 \pm 0.62$$

Thus we can be 95% confident that the true slope is between 0.76 and 2.00% per kg.

(g) Confidence limits for the long-term mean of y for a specified value of x are given by:

$$(a+bX) \pm t(\text{RSD})\sqrt{\{(1/m)+(1/n)+[(X-\bar{x})^2/\text{SXX}]\}}$$
$$= 3.25 \pm 2.57\ (0.6342)\sqrt{\{0+(1/7)+[(3.0-2.0)^2/7.0]\}}$$
$$= 3.25 \pm 0.87\%$$

(h) The calculated slope is 1.38% per kg. However, we must not forget that this is just an estimate. If you had produced more batches or fewer batches you would have obtained a different estimate, almost certainly. Thus, simply to quote the 1.38 in your report would be unhelpful to the reader, because you would be giving no indication of the precision of the estimate. If you present the 95% confidence interval as 1.38±0.62 or as 0.76 to 2.00 you convey not only the estimate but also the uncertainty. Similarly, the predicted impurity of 3.25% in part (g) is also an estimate. By reporting 3.25±0.87 you give some indication of the possible error in the estimate. Thus the reader can see that the mean impurity is likely to be between 2.38% and 4.12%.

3.11 DETAILED OBJECTIVES FOR THIS CHAPTER

Now that you have studied the material in this chapter and related it to your knowledge of previous chapters, you should be able to do the following.

(1) Explain the meanings of the following terms and use them correctly in appropriate contexts:
   (a) Scatter diagram;
   (b) best straight line or regression line;
   (c) regression equation;
   (d) calculated slope and true slope;
   (e) calculated intercept and true intercept;
   (f) independent variable, or predictor variable;
   (g) dependent variable or response variable;
   (h) correlation coefficient;
   (i) percentage fit;
   (j) residual standard deviation;
   (k) confidence limits for the true slope and the true intercept;
   (l) confidence limits for the true value of one variable corresponding to a particular value of the other variable.

(2) Decide which of the two variables can best be regarded as the independent variable.
(3) Use your calculator to obtain the slope, the intercept and the correlation coefficient.
(4) Explain the assumptions underlying simple regression analysis.
(5) Calculate residuals and use them to check the assumptions.
(6) Explain the dangers associated with the use of the correlation coefficient.
(7) Use Table ST3 to check the significance of a calculated correlation coefficient.
(8) Calculate the residual standard deviation.
(9) Calculate confidence limits for the true slope and the true intercept.
(10) Calculate confidence limits for the true value of one variable corresponding to a specified value of the other.

3.12   SELF-TEST

A well-known builder, in conjunction with Eurogas, is investigating the efficiency of a new type of cavity insulation. Eight householders, with identical houses on a particular estate, are persuaded to have the system fitted. The gas consumption in each house is recorded for the 12 months preceding and 12 months following the installation (Table 3.9).

Table 3.9 — Gas consumption in eight houses (therms)

House	A	B	C	D	E	F	G	H
Consumption before	339	427	318	385	262	320	410	306
Consumption after	322	392	246	304	279	273	336	294

Questions (1)–(10) refer to the data in Table 3.9. You might wish to check your answers to earlier questions before attempting the later ones.

(1) Which of the two variables would you choose as the predictor variable (x)?
   (a) 'consumption after', because the researcher would want to predict 'consumption before';
   (b) 'consumption after', because there is a direct cause and effect relationship between the two variables;
   (c) 'consumption before' because the researcher would want to predict 'consumption after';
   (d) 'consumption before' because in most houses it would be greater than 'consumption after'.
(2) What is the slope of the regression line?
   (a) 0.573
   (b) 0.844
   (c) 0.990
   (d) none of the above
(3) What is the intercept of the regression line?

  (a) 43.07
  (b) 71.55
  (c) 121.84
  (d) none of the above
(4) What meaning can you attach to the slope of the regression line?
  (a) Two houses which differ in consumption by 63 therms before installation
       could be expected to differ by 100 therms afterwards.
  (b) Two houses which differ in consumption by 100 therms before installation
       could be expected to differ by 63 therms afterwards.
  (c) A house could be expected to decrease its gas consumption by 63 therms as
       a result of installing the insulation.
  (d) A house could be expected to decrease its gas consumption to 63% as a
       result of installing the insulation.
(5) What meaning can you attach to the intercept of the regression line?
  (a) A house which used 89 therms before installation could be expected to have
       zero consumption afterwards.
  (b) A house which had zero consumption before installation could be expected
       to use 89 therms in the 12 months later.
  (c) Two houses which had the **same** gas consumption before installation could
       differ in consumption by as much as 89 therms after.
  (d) Two houses which differed in consumption by 89 therms before installation
       could be expected to have the **same** consumption afterwards.
(6) What is the residual standard deviation for the data in Table 3.9?
  (a) 19.92
  (b) 27.69
  (c) 29.91
  (d) 37.63
(7) From the data in Table 3.9 we could calculate confidence limits for the true
     slope. What is the meaning of 'true slope' in this situation?
  (a) It is the slope that we would have calculated if all eight houses had the same
       consumption **before** installation.
  (b) It is the slope that we would have calculated if all eight houses had the same
       consumption **after** installation.
  (c) It is the mean consumption for all houses **after** installation divided by the
       mean consumption **before** installation.
  (d) None of the above.
(8) Which of the following are 95% confidence limits for the mean consumption
     after installation for all houses which use 300 therms per year before
     installation?
  (a) 196 to 358 therms
  (b) 243 to 311 therms
  (c) 261 to 293 therms
  (d) none of the above.
(9) Which of the following statements is true concerning the correlation between
     the two variables in Table 3.9?
  (a) The calculated correlation (0.787) is greater than the required value
       (0.632) from Table ST3.

(b) The calculated correlation (0.641) is less than required value (0.707) from Table ST3.

(c) If the calculated correlation were greater than the required value from Table ST3 we would conclude that fitting the insulation tended to increase the gas consumption.

(d) none of the above.

(10) What meaning can we give to the residual standard deviation for the data in Table 3.9?

(a) It is the standard deviation of consumption that we would expect for those houses which did not change their consumption when insulation was fitted.

(b) It is the standard deviation of consumption that we would expect if all houses changed their consumption by the same amount when insulation was fitted.

(c) It is the standard deviation of consumption **after** fitting insulation that we would expect for houses which had the same consumption **before**.

(d) It is the standard deviation of consumption **before** fitting insulation that we would expect for houses which had the same consumption **after**.

3.13   ANSWERS TO SELF-TEST QUESTIONS

(1) (c)
(2) (d) (the slope is 0.625 78)
(3) (d) (the intercept is 89.3079)
(4) (b)
(5) (b)
(6) (c)
(7) (d)
(8) (b)
(9) (d)
(10) (c)

4

Significance testing

4.1 INTRODUCTION

In Chapter 3 we discussed regression and correlation. You utilized the full power of your calculator to obtain the slope and the intercept of the regression line, and then you assessed the fit of the line by means of the correlation coefficient. At the very end of Chapter 3 came the warning 'It is unwise to attach too much importance to a regression line unless the correlation coefficient is greater than the **required value** from Table ST3'. You may have wondered where these required values came from.

In Chapters 1 and 2 we gave some thought to **sampling**. I pointed out that the conclusions drawn from a set of data could be misleading if the sample was not representative of the population. Random sampling was put forward as a method which is likely to give a representative sample.

In this chapter we shall take another look at the correlation coefficient and its table of required values. We shall see that one of the assumptions underlying the table is concerned with how the sample was selected from the population. We shall also look at another table of required values which can be used as yardsticks when we wish to draw conclusions about population means.

The techniques of data analysis that we examine in this chapter have a similar purpose to the use of confidence intervals in Chapters 1, 2 and 3. This chapter will, therefore, help to consolidate your understanding of confidence intervals. In addition, these new techniques will also introduce you to some important ideas that have not been discussed in earlier chapters. The aims of this chapter are as follows.

(1) The primary aim of this chapter is to equip you to use simple significance tests. These are in widespread use and are alternatives to some of the techniques we used in Chapters 1 and 2.
(2) A second aim is to familiarize you with the philosophy and the assumptions that underlie these significance tests, so that you may use them safely and correctly.
(3) A further aim is to increase your appreciation of the risks associated with the statistical techniques used in this book. Whenever we use data from a sample to

draw conclusions about a population there is a danger that we will be misled, because the variability tends to obscure the effects which we are investigating.

4.2   THE CORRELATION TEST

We used a correlation coefficient in Chapter 3 to assess the 'goodness of fit' of a regression line. We were using the regression equation to make predictions and we realized at the time that these predictions would not be very reliable if the correlation coefficient were small. Let us take another look at correlation in a different context.

Ian Southwell is worried about day-to-day variations in the tensile strength of a synthetic yarn. This variation may have existed for many years but has only recently been identified, since the introduction of a two-shift system in the spinning shed. Previously, only one batch was made per day. Thus day-to-day variation was synonymous with batch-to-batch variation. After the introduction of two shifts it became possible to make two or even three batches per day and the day-to-day variation then became apparent.

Dr Southwell has long suspected that fluctuations in tensile strength could be due, at least in part, to changes in atmospheric conditions. He wonders therefore, whether the day-to-day variation could be explained in terms of the temperature and/ or the humidity within the spinning shed. These are recorded in the morning and afternoon for five consecutive working days (Table 4.1).

Table 4.1 — Ten batches of synthetic yarn

Day	1		2		3		4		5	
	am	pm	am	pm	am	pm	am	pm	am	pm
Tensile strength	87	92	74	71	83	80	87	89	80	77
Temperature (°C)	17	17	16	18	16	15	14	16	18	19
Humidity (%)	32	31	37	40	38	34	34	35	37	41

On each of the five days, two batches were produced, one in the morning and one in the afternoon. Thus it is possible to associate each tensile strength determination with a temperature and a humidity. Dr Southwell explores the data by drawing two scatter diagrams. One indicates that there **might** be a relationship between tensile strength and humidity. The other appears to offer no clear evidence of a relationship between tensile strength and temperature. Clearly Dr Southwell needs to check the correlation coefficients.

Exercise 4.1

(a) Use your calculator to obtain the correlation between tensile strength and humidity.

(b) Compare your calculated correlation coefficient from part (a) with the required value from Table ST3. What conclusion do you draw concerning the relationship between tensile strength and humidity?

What you did in Exercise 4.1 is, on the surface, very simple. However, if we inspect more closely the procedure and the conclusion, you will see that this is a widely applicable technique with very subtle implications. Consider first the wording of the conclusion: 'There is a relationship between tensile strength and the humidity within the spinning shed'. This does not apply only to the ten batches in Table 4.1. It applies to a whole population of batches. Our conclusion is a generalization based on a small set of data. The ten batches which gave us the data are a sample from a population. The need to draw conclusions about a population from sample data, arises so often that we have a name for the procedure — 'significance testing'.

A significance test is used
to draw a conclusion about a population,
using data from a sample.

Significance testing is so useful in data analysis that we adopt a formal procedure. At first sight the formality may appear unnecessary, or even offensive. However, after studying several types of significance test, you will probably regard the formal procedure as a great help, rather than a hindrance. I shall now repeat Exercise 4.1 using the recommended procedure, and you can judge for yourself.

The correlation test

Null hypothesis	There is no relationship between humidity and tensile strength.
Calculated value	= The sample correlation coefficient
	= 0.795
	(Note: when the calculated correlation is negative we ignore the minus sign).
Required value	= 0.632 (from Table ST3 for a sample size of 10).
Conclusion	Because the calculated value is greater than the required value we reject the null hypothesis and conclude that there is a relationship between tensile strength and humidity.

We shall discuss this four-step procedure in detail later in the chapter. For the moment I shall simply comment briefly on the philosophy on which it is based. The null hypothesis is a simple statement which might, or might not, be true. To make a decision about the truth of the hypothesis we compare the evidence (i.e. something calculated from the data) with a standard, drawn from a statistical table. If the evidence is not consistent with the hypothesis we reject the hypothesis and draw an appropriate conclusion.

Exercise 4.2

(a) Using the data in Table 4.1 calculate the correlation between tensile strength and temperature.
(b) Carry out a correlation test, following the formal procedure very carefully.

In Exercise 4.2 you were advised to follow the procedure carefully. Perhaps you felt that you were following the procedure blindly. Since you have not studied the mathematical foundations of Table ST3, it is understandable that you should feel uneasy about the whole procedure. I am confident, however, that you will find significance testing acceptable, after we have discussed the procedure in more detail. Before we do so, I should like to introduce a second type of test.

4.3 THE ONE-SAMPLE *t*-TEST

There are many types of significance test. We shall now turn our attention to a test known as the 'one-sample *t*-test'. Later we shall discuss the 'two-sample *t*-test'. All *t*-tests are concerned with **means** and have nothing to do with correlation. When carrying out a one-sample *t*-test we draw a conclusion about a population mean using, among other things, the mean of a sample. 'But', you might protest, 'didn't we achieve this in Chapter 1, using confidence limits?' Well, we achieved something similar. Let us take another look at it. The first set of data that we analysed in Chapter 1 contained the yield and impurity of six batches of digozo blue pigment. These data were based on six batches produced shortly after certain modifications to the process had been carried out. We used the data to calculate confidence limits for the mean yield and the mean impurity of future batches of digozo blue. The 95% confidence limits for the population mean yield were 661.8 kg and 734.2 kg. We therefore reported that the mean yield of future batches would lie between these two weights.

The above paragraph describes how we attempted to answer the question 'What will be the mean yield of future batches?' The plant manager is very satisfied with our answer, but he has an additional question concerning the effect of the recent modification. 'Has the mean yield decreased?', he wishes to know. The mean yield before the modification was approximately 725 kg (this figure was estimated from several months' production, after eliminating one or two suspect batches). The mean yield of the six batches produced since the modification is 698 kg.

Exercise 4.3

(a) In response to the question 'Has the yield decreased on average since the modification was carried out?', should we answer 'yes' or 'no'?
(b) As your thinking progressed from the data to your 'yes' or 'no' answer, what did you take into account?

Let us now use a one-sample t-test to answer the question posed in Exercise 4.3(a). This test will follow the four-step procedure that we used in the correlation test. The calculated value will be based on the sample mean, the sample standard deviation, the sample size and a number which is obtained from the null hypothesis.

The one-sample t-test

Null hypothesis	The mean yield has not changed (in other words, the population mean yield is equal to 725 kg).
Calculated value	$= (|\bar{x} - \mu|\sqrt{n})/s$
	$= (|698{-}725|\sqrt{6})/34.50$
	$= 1.92$
	where $\bar{x}$ is the sample mean, s is a suitable standard deviation, n is the size of the sample from which the mean was calculated and μ is the population mean (the value of μ is not known, but is given by the null hypothesis).
Required value	$= 2.57$ (from the 95% column of Table ST1, with 5 degrees of freedom).
Conclusion	As the calculated value is less than the required value we cannot reject the null hypothesis. Thus we are unable to conclude that the mean yield has changed.

Obviously there is a strong similarity between the formula used in the one-sample t-test and the formula used to calculate confidence limits for a population mean. The two formulae are equivalent, in fact. Thus, whenever the confidence interval includes the specified value, we will find that the calculated value is less than the required value. It is not surprising, therefore, that the two procedures always lead to the same conclusion.

Exercise 4.4

Ten lumps of coal are taken from a 100 tonne consignment and a determination of sulphur content is made on each. The mean and standard deviation of these ten determinations are 0.341% and 0.041 22%.

(a) Carry out a one-sample t-test to see whether the mean sulphur content of the consignment differs significantly from the 0.31% given in the supplier's certificate of analaysis. (Hint: if you follow the worked example carefully this is an easy task. However, of you deviate from the example you may tie yourself in knots.)
(b) Can you think of any reason why the conclusion you drew in part (a) could be misleading?

As an alternative to the one-sample t-test in Exercise 4.4. you could have calculated 95% confidence limits for the population mean, then asked 'Does the confidence interval include the certified value of 0.31%?' If you calculate the confidence limits you will find that they are 0.312% and 0.370%. Thus the interval does not include 0.31% and we would, therefore, conclude that the true sulphur content of the consignment was not equal to the certified value. Once again the two alternative techniques have led us to the same conclusion as they always will.

4.4 THE SIGNIFICANCE TEST PROCEDURE

Already we have carried out two types of significance test, the correlation test and the one-sample t-test. Later in this chapter we shall examine other types of test. Before we do so, however, I would like to comment on what we have achieved so far and to list some featuers which are common to all significance tests. Perhaps I could start by repeating the definition.

A significance test is used
to draw a conclusion about a **population**
using data from a **sample**

Read the above definition again — and again. Obviously, significance testing only makes sense when we consider the sample **and** the population. It is the **population** about which we wish to draw a conclusion, but we cannot examine the whole population, so we base our conclusion on data from a **sample**. Clearly significance testing is dangerous.

The conclusion resulting from a significance test could be misleading if the sample is not representative of the population. You may recall, from your study of Chapter 1, that random sampling is often used in an attempt to obtain a representative sample.

An assumption
underlying **all** significance tests
is that random sampling was used.

In many situations it is not possible to use random sampling, so some other method of sampling is employed. Whatever sampling method you use the assumption of random sampling remains. It is built into all the statistical tables. If you are unable or unwilling to use random sampling you should ask yourself 'Is my sample likely to be representative of the population?' before proceeding with a significance test. If the answer is a clear 'no' then your conclusions could be misleading, no matter how sophisticated your analysis. Note that confidence intervals are just as dangerous because they share the random sampling assumption.

Some significance tests have additional assumptions. For example, the one-sample t-test is based on the second assumption that the population has a **normal distribution**. In practice it is never possible to prove that this assumption is satisfied, but we can inspect the data to see whether it is obviously violated. Fortunately the normal distribution assumption is less important if your sample is large. With a sample of 30 or more you need not worry at all. Of course, it is always wise to carry out a graphical inspection of data before subjecting it to any statistical analysis. A picture is more likely to reveal any abnormality than is an inspection of the numbers.

Let us now turn our attention to the significance testing procedure. Perhaps this will seem more reasonable if I comment briefly on each of the four steps.

Step 1: The Null Hypothesis

A hypothesis is a statement which might, or might not, be true. In any significance test we use the data to decide whether or not the null hypothesis is true. The null hypothesis always refers to the **population**, not the sample. Indeed the null hypothesis could be written down **before** we took the sample. The null hypothesis can be written as an equation. For example, in the correlation test, the null hypothesis states 'The population correlation is equal to zero'. In the one-sample t-test the null hypothesis might be 'The population mean is **equal** to 0.31%'.

Step 2: The Calculated Value

The calculated value is based on the data, of course, and cannot be obtained until the experiment or the survey is complete. Usually the data will be summarized by calculating a mean, a standard deviation, a correlation or a percentage; then these may be substituted into a formula to work out the calculated value. The formula may contain both English letters and Greek letters. These will always conform to the convention that English letters refer to the sample while Greek letters refer to the population. For example $\bar{x}$ represents the sample mean and μ represents the population mean, in the formula for the one-sample t-test. In practice the numbers which replace the English letters will come from the data but the numbers which replace the Greek letters will come from the null hypothesis. If the null hypothesis is true we would expect the calculated value to be small. Thus whenever we obtain a large calculated value we know that:

(a) the null hypothesis is **not** true, or
(b) the null hypothesis **is** true, but we have drawn a sample which is not representative of the population, or
(c) one of the assumptions underlying the significance test is not satisfied.

Thus, if we can rule out (b) and (c), we can happily reject the null hypothesis whenever we obtain a large calculated value. 'But how am I to know what is large and what is small?', you might wonder. Read on.

Step 3: The Required Value

We reject the null hypothesis if the calculated value is greater than the required value taken from the appropriate statistical table. The required value is acting as a yardstick against which we compare the data. We take the required value from the 95% column of the table. Thus there is 95% chance that the calculated value will be less than the required value, if the null hypothesis is true. Hence there is only 5% chance of rejecting a null hypothesis which is true. Many users of significance tests prefer the phrase '5% significance level' to the phrase we used in earlier chapters, '95% confidence level'.

Step 4: The Conclusion
If the calculated value is **greater** than the required value we reject the null hypothesis and draw an appropriate conclusion. In the correlation test we would conclude that the population correlation was **not equal** to zero. In the one-sample *t*-test we would conclude that the population mean was **not equal** to the value put forward in the null hypothesis.

If, however, the calculated value is **less** than the required value we cannot reject the null hypothesis. Our conclusion is then expressed as a double negative, e.g. 'We are unable to conclude that the population correlation is not equal to zero'. You may not like this terminology. Please note, however, that **we never accept the null hypothesis**. It would be illogical to do so. The required values in the statistical table are based on the presumption that the null hypothesis is true. You cannot presume that something is true in order to prove that it is true.

> We may prove, beyond reasonable doubt,
> that the null hypothesis is false,
> but we cannot prove that it is true.

Perhaps this line of reasoning will become more acceptable if we compare our significance testing procedure with the procedure followed in English criminal courts. Just as the defendant is assumed innocent until proved guilty we assume that the null hypothesis is true until it is proved false. Absolute proof is not expected with either procedure; proof beyond reasonable doubt is an acceptable compromise. The two possible court verdicts 'guilty' and 'not guilty' mirror the two possible decisions in a significance test, 'reject the null hypothesis' and 'cannot reject the null hypothesis'. We never accept the null hypothesis just as the courts never declare a defendant innocent.

This discussion of the ideas underlying significance testing has been rather abstract. You may have found it difficult to absorb. Let us now return to more practical matters as we explore the use of two other types of significance test. I am sure you will find this easier. However, before we return to earth I would like to give a little more thought to the philosophy of significance testing.

Exercise 4.5

I would like you to insert the missing words in the following passage. You may need to think quite deeply and you may need to refer back to the text, as you are unlikely to have fully understood and memorized the discussion of significance testing.

A significance test is carried out in order to draw a conclusion about a (a) _____ using data from a sample. However, the conclusion could be misleading if the sample is not (b) _____ of the population. An assumption underlying every significance test is that (c) _____ sampling was used to select the sample from the population. The significance testing procedure consists of (d) _____ steps. The first step is to write down the (e) _____ (f) _____ .
This could be done **before** the sample is taken since the hypothesis never

refers to the (g) _____ , but always refers to the population. To reach a decision about the truth of the hypothesis we use the data to calculate a (h) _____ (i) _____ and compare this with a (j) _____ (k) _____ obtained from a statistical table. If the (l) _____ value is greater than the (m) _____ value we (n) _____ the null hypothesis. Thus we have proved beyond reasonable doubt that the hypothesis is (o) _____ . The required value is taken from the 95% column of the statistical table. Thus there is only (p) _____ chance of rejecting the null hypothesis if it is true. For this reason many users of significance tests refer to the 5% (q) _____ (r) _____ rather than the 95% confidence level.

4.5 THE TWO-SAMPLE *t*-TEST

Dr Protus has conducted an experiment to compare two alternative pig feeds. 15 pigs of the same age, but each from a different litter, were assigned at random to either feed A or feed B. The weights of the pigs at the end of the 30 day trial are given in Table 4.2.

Table 4.2 — Weights of 15 pigs on two diets

Feed	Weight (kg)	Mean	SD
A	22.4, 26.6, 18.8, 20.2, 19.6, 24.5, 28.0, 19.9	22.50	3.4897
B	27.6, 30.0, 24.5, 30.2, 22.6, 28.7, 24.0	26.80	3.0773

To make a judgment on the two types of feed Dr Protus compares the two means in Table 4.2. He concludes that feed B is more effective than A, simply because 26.8 kg is greater than 22.5 kg. This conclusion may be correct. However, if Dr Protus continues to use this method of decision making he is destined to draw many conclusions which are **not** correct.

Suppose Dr Protus asked **you** to analyse his data. You would compare the two means of course, but you would also take account of the sample sizes and of the variability in weight from pig to pig. Having studied Chapter 2, you might achieve this by calculating confidence limits for the difference between two population means, then asking 'Does the confidence interval include zero?' The difference between the sample means is 4.30 kg, the combined standard deviation is 3.305 and the calculation gives confidence limits equal to 0.61 kg and 7.98 kg. Thus we can be 95% confident that the extra weight gain to be achieved by using feed B rather than A lies between 0.61 kg and 7.98 kg. The lower confidence limit is small, of course, but it is greater than zero, so we can claim to have proved beyond reasonable doubt that feed B is superior to A.

The analysis carried out in the above paragraph contains nothing new. We did exactly the same in Chapter 2. Let is now progress to an alternative method of analysis, known as the two-sample *t*-test, which many people would prefer. This test uses exactly the same ingredients, the sample means, the combined standard deviation and the sample sizes. Not surprisingly, it will lead us to exactly the same conclusion.

The two-sample *t*-test

Null hypothesis	The two diets are equally effective.
Calculated value	$= (\bar{x}_1 - \bar{x}_2)/\{s\sqrt{[1/n_1) + (1/n_2)]}\}$
	$= (26.80 - 22.50)/\{3.305\sqrt{[(1/7) + (1/8)]}\}$
	$= 2.51$
	where $\bar{x}_1$ is the larger of the two sample means, $\bar{x}_2$ is the smaller of the two sample means, s is a suitable standard deviation (probably the combined standard deviation for the two samples) and n_1 and n_2 are the sizes of the two samples from which the means were calculated.
Required value	$= 2.16$ (from the 95% column of Table ST1, with the same degrees of freedom as the standard deviation, i.e. 13).
Conclusion	As the calculated value is greater than the required value we reject the null hypothesis and conclude that the two feeds are **not** equally effective.

As I predicted, the two-sample *t*-test has led us to the same conclusion that we reached when we calculated confidence limits for the difference between the two population means. This will be the case with any set of data. Thus, you may use whichever of the two techniques appeals to you most. Obviously the two-sample *t*-test is just as useful as the confidence interval alternative, which we used extensively in Chapter 2. In fact the two-sample *t*-test is so widely used that I would like to consider a second example of its application.

Dr Lubrick is a research chemist employed by the Chugwell Oil Company. He is attempting to develop an additive which, when included in engine oil, will decrease petrol consumption. A trial is arranged in which 12 cars are driven for 20 000 miles during which time their petrol consumption is recorded. Six of the cars have oil containing the additive while the other six use normal oil (Table 4.3).

The mean petrol consumption for the six cars with additive is 39.25 and the mean for the six cars without is 34.40. The experiment indicates, therefore, that the use of the additive offers an increase in miles per gallon of 4.85. Dr Lubrick is excited by this finding. He realizes that many motorists would gladly pay a little more for a special engine oil, if they believed that they could get almost 5 mpg more than with normal oil.

Table 4.3 — Petrol consumption (mpg) of 12 cars

With additive						
Car	A	B	C	D	E	F
Consumption	23.7	46.8	32.1	52.7	45.2	35.0
Without additive						
Car	G	H	I	J	K	L
Consumption	48.2	21.4	36.2	42.7	26.5	30.4

Exercise 4.6

(a) Calculate the combined standard deviation for the two sets of data in Table 4.3.
(b) Carry out a two-sample *t*-test to see whether the observed increase in mpg is statistically significant.
(c) Dr Lubrick is prepared to carry out a second experiment in an attempt to prove beyond reasonable doubt that the use of the additive increases the miles per gallon. How should the second experiment differ from the first, in order to increase the chance of revealing the effectiveness of the additive?

4.6   THE PAIRED COMPARISON TEST

Let us consider the oil additive experiment in more detail. Dr Lubrick assigned the cars at random to the two treatments. So each car had a 50:50 chance of getting oil containing the additive. You might wonder what type of cars were used in the experiment and how they were selected? Dr Lubrick deliberately included a wide variety of cars so that he could prove that his additive was effective in **any** make of car. The car which gave only 21.4 mpg was very large whereas the car which gave 52.7 mpg was much smaller.

Therefore the enormous variability in petrol consumption from car to car was actually introduced by Dr Lubrick. Obviously the standard deviations would have been smaller if he had used 12 cars of the same type, and with much smaller standard deviations the difference between the sample means might have been statistically significant. Of course, if Dr Lubrick used only **one** type of car he could not reasonably claim that his additive was effective with **every** make of car.

There are two solutions to this problem. The first alternative is to extend the experiment to include hundreds of cars. We would then find that a difference in means of 5 mpg would be statistically significant even with such a large standard deviation. The second, and much less expensive, alternative is to use pairs of 'identical' cars. If Dr Lubrick took this second course of action he might well get data similar to those in Table 4.4.

The numbers in Table 4.4 are exactly the same as those in Table 4.3. Of course, they take on a totally different meaning in this hypothetical experiment and they require a very different analysis. The important point to note about Table 4.4 is that

Table 4.4 — Petrol consumption (mpg) of 12 cars

Type of car	U	V	W	X	Y	Z
With additive	23.7	32.1	35.0	45.2	46.8	52.7
Without additive	21.4	26.5	30.4	36.2	43.7	48.2

the numbers can be **paired**. The 23.7 and the 21.4 are petrol consumption figures for two 'identical' large cars. Similarly the 52.7 and the 48.2 are from two 'identical' small cars. The hypothetical experiment which gave us these data can be described as a paired comparison experiment and the method of analysis we shall use is often referred to as the paired comparison test. It is helpful to break down the analysis into two steps:

(a) calculate the difference for each pair;
(b) do a one-sample t-test on the differences.

Now that we have the **differences** in Table 4.5 we shall make no further use of the

Table 4.5 — Differences in consumption for six pairs of cars

Type of car	U	V	W	X	Y	Z	Mean	SD
With additive	23.7	32.1	35.0	45.2	46.8	52.7	39.25	10.810
Without additive	21.4	26.5	30.4	36.2	43.7	48.2	34.40	10.272
Difference	2.3	5.6	4.6	9.0	3.1	4.5	4.85	2.467

petrol consumption figures. The mean and standard deviation of the differences will be used to calculate the test value in our paired comparison test.

The calculated value in a paired comparison test is obtained from the following formula:

$$\text{calculated value} = (|\bar{x} - \mu| \sqrt{n})s$$

where $\bar{x}$ is the mean of the differences, s is the standard deviation of the differences, n is the number of differences and μ is equal to zero.

The formula above is exactly the same as that used in the one-sample t-test. With the paired comparison test, however, $\bar{x}$, s and n are based on the **differences**. Thus a paired comparison test is simply a one-sample t-test based on the differences between pairs of measurements. Taking advantage of the pairing can have a dramatic effect, however.

The paired comparison test

Null hypothesis	The population mean difference is equal to zero.
Calculated value	$= (|\bar{x} - \mu|\sqrt{n})/s$
	$= (|4.85 - 0|\sqrt{6})/2.3467$
	$= 5.06$
Required value	$= 2.57$ (from Table ST1 with 5 degrees of freedom and 95% confidence).
Conclusion	Because the test value is greater than the table value we reject the null hypothesis and we conclude that the use of the additive does reduce petrol consumption.

The paired comparison test leads us very strongly to the conclusion that the additive is effective, whereas the two-sample *t*-test indicated that the use of the additive made no significant difference. The extra power of a paired comparison test comes from its use of the **standard deviation of the differences**. If we look back at Table 4.5 we see that this standard deviation (2.3467) is much smaller than the other two. The values 10.810 and 10.272 are so large because of the enormous variation in consumption from the small cars to the large cars. By calculating the differences, however, we eliminate much of this variability. This point has serious implications for any scientist who is designing an experiment.

It is worthwhile to give further consideration to the standard deviations in Table 4.5. The petrol consumption figures in the table contain a great deal of variability. If we type all 12 numbers into our calculator we find a standard deviation equal to 10.368 mpg. It can reasonably be argued that this variability is partly due to each of the following:

(a) variability in engine and efficiency from one model to another;
(b) variability from car to car of the same model;
(c) variability in many extraneous factors such as the performance of the drivers and the weather;
(d) the effect of the additive.

We are trying to assess (d), but it is hidden by (a), (b) and (c). However, (a) is much bigger than (b) or (c) so if we eliminate (a) we stand a much better chance of finding (d). By calculating differences in Table 4.5 we automatically remove the variability from model to model and our standard deviation falls from 10.38 to 2.3467. When we compare our mean difference (4.85) with this reduced standard deviation in the paired comparison test it proves to be significant. This demonstrates the wisdom of using six pairs of 'identical' cars (i.e. two Metros, two Sierras, etc.).

Using the additional power of the paired comparison test we have proved beyond doubt that the additive is effective in reducing petrol consumption. 'How effective?', you may ask, 'If Dr Lubrick persuades his company to introduce the additive into engine oil how large an increase in miles per gallon can he expect to achieve?' We can calculate confidence limits for the increase in mpg if we adapt a formula introduced in Chapter 1.

95% confidence limits for the difference between two treatments in a paired comparison experiment are given by:

$$\bar{x} \pm ts/\sqrt{n}$$

where $\bar{x}$ is the mean difference, s is the standard deviation of the differences, t is taken from Table ST1 with $(n-1)$ degrees of freedom and n is the number of differences.

Exercise 4.7

A classic example of the paired comparison experiment is provided by an assessment of the effectiveness of a diet. The data in Table 4.6 give the weights, before and after

Table 4.6 — Data from a dietary experiment

Person	A	B	C	D	E	F	G	H
Weight before (lb)	137	196	103	162	154	122	142	136
Weight after (lb)	130	190	98	153	156	115	143	133

dieting, of eight volunteers.

The mean weight before the diet is 144 lb and the mean weight after the diet is 139.75 lb. Thus the participants have lost 4.25 lb on average. If we were to carry out a two-sample t-test on these data we would find that the two sample means were not significantly different. However, this is not the best method of analysis, for there is a natural pairing in the data. There are two measurements on each person. You are asked, therefore, to carry out a paired comparison test.

(a) Calculate the weight loss for each person (note that two of the weight losses are negative).
(b) Calculate the mean and standard deviation of the eight weight losses (use the $+/-$ key to cope with the negative values).
(c) Carry out a one-sample t-test on the weight losses and draw any conclusion about the effectiveness of the diet.
(d) Calculate 95% confidence limits for the mean weight loss that can be expected with this diet.

4.7 SUMMARY OF THE IMPORTANT POINTS

(1) In this chapter we have reconsidered several sets of data which were introduced in earlier chapters. Previously we had analysed these data using confidence intervals but in this chapter we have used **significance tests**.

(2) We have used the **correlation test** to check the relationship between two variables.
(3) We have used the **one-sample *t*-test** to compare a treatment with an external reference value.
(4) We have used the **two-sample *t*-test** to compare two treatments.
(5) We have also used the **paired comparison test** to compare two treatments, but this analysis is only appropriate when a paired comparison experiment has been carried out.
(6) These four significance tests are summarized in Table 4.7.

Table 4.7 — Significance tests introduced in this chapter

Type of test	Null hypothesis	Test value	Table value
Correlation test	The population correlation is equal to zero	The sample correlation coefficient	From Table ST3
One-sample *t*-test	The population mean is equal to some specified value	$(\lvert \bar{x} - \mu \rvert \sqrt{n})/s$	From Table ST1 with degrees of freedom appropriate to the standard deviation
Two-sample *t*-test	The two population means are equal	See below	From Table ST1 with degrees of freedom appropriate to the standard deviation
Paired comparison test	The population mean difference is equal to zero	$(\lvert \bar{x} - \mu \rvert \sqrt{n})/s$	From table ST1 with $n - 1$ degrees of freedom

In the two-sample *t*-test the calculated value is equal to $(\bar{x}_1 - \bar{x}_2)/\{s\sqrt{[(1/n_1) + (1/n_2)]}\}$

4.8 ADDITIONAL EXERCISES

Exercise 4.8

Dr Maxwell is conducting a clinical trial to evaluate the effectiveness of a new drug. It is hoped that this drug will reduce the pulse rate of patients suffering from a particular disease. 20 patients suffering from this disease are treated with the drug for 12 weeks, their pulse rates being measured both before and after the treatment. The results of the trial are given in Table 4.8.

In this chapter we have discussed three types of significance test. Which type of test would you use to answer each of the following questions?

(a) Do female patients have the same pulse rate on average as male patients before the drug is taken?
(b) Do female patients have higher pulse rate on average than male patients after the drug has been taken over 12 weeks?
(c) Does the drug reduce the pulse rate of male patients?
(d) Does the mean pulse rate of the patients after treatment differ from the mean pulse rate for all people of this age, which is 68 beats per minute?
(e) (More difficult) Is the drug more effective with male patients than with female patients?

Check your answers to this question before proceeding with Exercise 4.9.

Exercise 4.9

Using the data from Exercise 4.8 carry out appropriate significance tests to answer the following questions.

(a) Can we reasonably conclude that the drug is effective in reducing the pulse rate of male patients suffering from this disease?
(b) Is the mean pulse rate of female patients suffering from this disease greater than the mean of all females of this age, which is 68 beats per minute?
(c) Is the mean pulse rate of male patients suffering from this disease less than the mean for all males of this age, which is 68 beats per minute? (Take extra care. Significance testing is not a substitute for common sense.)
(d) Do male and female sufferers from this disease have equal pulse rates on average after treatment with this drug?

4.9   WORKED SOLUTIONS

Solution to Exercise 4.1

(a) The calculated correlation coefficient is equal to -0.795 for the humidity and tensile strength data.
(b) With a sample size of 10 the required value from Table ST3 is 0.632. As the calculated correlation is greater than the required value we can conclude that there is a relationship betwen tensile strength and the humidity in the spinning shed.

Table 4.8 — Pulse rate (beats per minute) of 20 patients

Person	A	B	C	D	E	F	G	H	I	J
Sex	M	F	M	M	F	M	F	F	F	M
Pulse rate before	85	96	81	63	70	84	77	62	70	60
Pulse rate after	78	91	78	65	66	82	79	54	65	62

Person	K	L	M	N	O	P	Q	R	S	T
Sex	F	F	M	F	M	F	F	M	F	F
Pulse rate before	62	67	77	54	86	74	79	96	64	65
Pulse rate after	68	64	71	51	80	72	85	90	60	67

Solution to Exercise 4.2

(a) The correlation between tensile strength and temperature is equal to 0.368.
(b) *Null hypothesis*   There is no relationship between temperature and tensile strength.
   Calculated value = 0.368.
   Required value   = 0.632 from Table ST3.
   Conclusion   As the calculated value is less than the required value we cannot reject the null hypothesis. Thus we cannot conclude that there is a relationship between temperature and tensile strength.

Note the wording of the conclusion. It is not reasonable to conclude that 'there is no relationship between temperature and tensile strength'.

Solution to Exercise 4.3

(a) As the sample mean of 698 kg is less than the specified weight of 725 kg, there is some evidence that the yield has decreased. Is this evidence conclusive? Since the confidence interval, 661.8 to 734.2 kg, contains 725 kg, you may well have concluded that the mean yield could still be the same as it was before the modification.
   You were asked to answer 'yes' or 'no', but you may be reluctant to choose either alternative. You do not wish to answer 'yes' because the evidence is not strong enough. However, you do not wish to answer 'no', because you cannot prove that the mean has not changed.
(b) In your deliberations you may have considered the confidence limits for the new mean, 661.8 and 734.2 kg, in relation to the old mean of 725 kg. Thus, indirectly, you will have taken account of the batch-to-batch variation in yield and the sample size. You will also be aware that the calculation of the confidence limits required a value of t from Table ST1, which has certain built-in assumptions.

Solution to Exercise 4.4

(a) *Null hypothesis* The mean sulphur content of the consignment in equal to 0.31%.

Calculated value $= (|\bar{x} - \mu|\sqrt{n})/s$

$= (|0.341 - 0.31|\sqrt{10})/(0.04122)$

$= 2.38$

Required value $= 2.26$, from the 95% column of Table ST1, with 9 degrees of freedom.

Conclusion As the calculated value exceeds the required value we can reject the null hypothesis and conclude that the sulphur content of this consignment is not 0.31%.

(b) The conclusion reached in part (a) could be misleading if any of the assumptions underlying this technique are violated. The assumptions are exactly the same as those discussed in Chapter 1 when we calculated confidence limits for the population mean. They are:

(i) the ten lumps of coal were selected at random from all lumps within the consignment;

(ii) the population of sulphur determinations would have a normal distribution.

Solution to Exercise 4.5

You may have found this exercise rather difficult. However, I hope you have tried very hard and re-read the text, several times if necessary. My answers are:

(a) population
(b) representative
(c) random
(d) four
(e) null
(f) hypothesis
(g) sample
(h) calculated
(i) value

(j) required
(k) value
(l) calculated
(m) required
(n) reject
(o) false
(p) 5%
(q) significance
(r) level

Solution to Exercise 4.6

(a) The standard deviations of the two sets of data in Table 4.3 are 10.8101 and 10.2721 mpg. The combined standard deviation is 10.5445.

(b) *Null hypothesis* The mean consumption with the additive is equal to the mean consumption without the additive (i.e. the additive has no effect).

Calculated value $= (\bar{x}_1 - \bar{x}_2)/\{s\sqrt{[(1/n_1) + (1/n_2)]}\}$

$= (39.25 - 34.40)/\sqrt{\{10.5445\,[(1/6) + (1/6)]\}}$

$= 0.80$

Required value $= 2.23$ from the 95% column of Table ST1 with 10 degrees of freedom.

Conclusion As the calculated value is less than the required value we

cannot reject the null hypothesis. Thus we are unable to
conclude that the additive is effective.

(c) The conclusion you reached in part (b) is a big disappointment for Dr Lubrick.
The difference of 4.85 mpg between the two sample means is sufficiently large to
be very important. However, it is **not** large enough to be statistically significant,
with such a large standard deviation and with such small samples. If the second
experiment is likely to give an equally large standard deviation, then the sample
sizes will need to be much larger. In practical terms, Dr Lubrick must either
increase the number of cars or decrease the variation in consumption from car to
car. However, he wishes to use a wide variety of cars so that he can claim that the
additive is effective with all types.

There is an easy solution to Dr Lubrick's problem. He should carry out a
different type of experiment which effectively eliminates the variation from car
to car.

Solution to Exercise 4.7

(a) The weight losses are 7, 6, 5, 9, -2, 7, -1 and 3 lb.

(b) Mean weight lost $= 4.25$ lb; SD of weight losses $= 3.955$ lb.

(c) *Null hypothesis*   The population mean difference is equal to zero (i.e. the diet
has no effect on average).

Calculated value $= (|\bar{x} - \mu|\sqrt{n})/s$
$$= (|4.25 - 0|\sqrt{8})/3.955$$
$$= 3.04$$

Required value $= 2.36$ ... from Table ST1 with 7 degrees of
freedom and 95% confidence.

Conclusion   Because the calculated value is greater than the required value
we reject the null hypothesis. We therefore conclude that the
diet is effective.

Note: the standard deviation of the weight losses (3.955) is much smaller than the
standard deviations of the weights before (27.800) and the weights after
(27.958). Calculation of the differences eliminates the enormous variation in
weight from person to person. It is this variability which swamps the difference
between the means in the two-sample t-test.

(d) 95% confidence limits for the mean weight loss are given by $\bar{x} \pm ts/\sqrt{n}$. Using $\bar{x} = 4.25$, $s = 3.955$, $n = 8$ and $t = 2.36$ we find

$$4.25 \pm (2.36)\,(3.955)/\sqrt{8}$$
$$= 4.25 \pm 3.31$$
$$= 0.94 \text{ to } 7.56 \text{ lb}$$

Thus we can be 95% confident that the mean weight loss for people on this diet
will lie between 0.94 and 7.56 lb. Of course this interval would have been
narrower if more than eight people had taken part in the experiment.

Solution to Exercise 4.8
(a) To answer this question we would use a two-sample t-test.
(b) This question, although it is worded rather differently to (a), also requires a two-sample t-test.
(c) A paired comparison test is required.
(d) A one-sample t-test would be used to answer this question.
(e) To answer this question we would calculate 'decrease in pulse rate' for every patient, and then carry out a two-sample t-test to compare the mean decrease for males with the mean decrease for females.

Solution to Exercise 4.9
(a) We shall carry out a paired comparison test on the data for male patients. First we must calculate the decrease in blood pressure for each patient.

Table 4.9 — Decreases in blood pressure for male patients

Patient	A	C	D	F	J	M	O	R
Before	85	81	63	84	60	77	86	90
After	78	78	65	82	52	71	80	90
Decrease	7	3	-2	2	8	6	6	6

The mean decrease = 4.5 beats per minute and the SD of decreases = 3.2950 beats per minute.

Null hypothesis   The population mean decrease is equal to zero (i.e. the drug has no effect on pulse rate).

Calculated value $= (|\bar{x} - \mu|\sqrt{n})s$
$= (|4.5 - 0|\sqrt{8})/3.2950$
$= 3.86$

Required value $= 2.36$ from Table ST1 with 7 degrees of freedom and 95% confidence.

Conclusion   Because the calculated value is greater than the required value we reject the null hypothesis and conclude that the drug does reduce the pulse rate of male patients.

(b) We shall carry out a one-sample t-test using the pulse rate figures for females before treatment.

Null hypothesis   The population mean is equal to 68 beats per minute.

Calculated value $= (|\bar{x} - \mu|\sqrt{n})s$
$= (|70.0 - 68|\sqrt{12})/10.7703$
$= 0.64$

Required value = 2.20 from Table ST1 with 11 degrees of freedom and 95% confidence.

Conclusion Because the calculated value is less than the required value we cannot reject the null hypothesis. Thus we are unable to conclude that female sufferers from this disease have higher pulse rate on average than other females.

(c) The mean pulse rate of the eight male patients before treatment was 79.0 beats per minute. Obviously this is **greater** than 68 beats per minute. Thus the evidence from the sample suggests that the mean pulse rate of all male sufferers form this disease is **greater** than 68. We could use a one-sample *t*-test to see whether this evidence is statistically significant. However, we are asked the question 'Is the mean pulse rate of males **less than** 68?' A sample mean of 79, which is **greater** than 68, cannot be used to prove that the population mean is **less** than 68. Thus it is pointless to carry out a significance test.

(d) Using the pulse rates after treatment we obtain:

$$
\begin{array}{llll}
\text{male:} & \text{mean} = 74.5 & \text{SD} = 11.7018 & n = 8 \\
\text{female:} & \text{mean} = 68.5 & \text{SD} = 11.7975 & n = 12 \\
\end{array}
$$
$$\text{combined standard deviation} = 11.7604$$

Null hypothesis The two population means are equal (i.e. male and female sufferers have the same pulse rate on average).

Calculated value $= (\bar{x}_1 - \bar{x}_2)/\{s\sqrt{[(1/n_1) + (1/n_2)]}\}$
$= (74.5 - 68.5)/\{11.7604\sqrt{[(1/8) + (1/12)]}\}$
$= 1.12$

Required value = 2.10 from Table ST1 with 18 degrees of freedom and 95% confidence.

Conclusion Because the calculated value is less than the required value we cannot reject the null hypothesis. Thus we are unable to conclude that male and female sufferers have different pulse rates on average.

4.10 DETAILED OBJECTIVES FOR THIS CHAPTER

Now that you have studied this chapter and attempted to relate its content to your previous knowledge, you should be able to do the following.

(1) Explain the meaning of the following terms and use them appropriately in suitable contexts:

(a) significance test;
(b) correlation test;
(c) one-sample *t*-test;
(d) two-sample *t*-test;
(e) paired comparison test;
(f) null hypothesis;
(g) calculated value;

  (h) required value;
  (i) statistically significant.

(2) Carry out a correlation test.
(3) Carry out a one-sample t-test.
(4) Carry out a two-sample t-test.
(5) Carry out a paired comparison test.
(6) Decided which of the above tests, if any, is appropriate in a specified situation.
(7) Explain the similarity between the four-step procedure used in a significance test and the procedure used in English criminal courts.
(8) Calculate confidence limits for the population mean difference in a situation where the paired comparison test is appropriate.
(9) Explain why the paired comparison test is more powerful than the two-sample t-test.

4.11 SELF-TEST

Dr Fayle is investigating the effect of alcohol on the speed of reaction of car drivers. The 12 subjects participating in his experiment are asked to 'drive' a simulated vehicle and their times of reaction to certain incidents are measured. Six of the subjects have drunk a controlled quantity of alcohol 10 minutes before the drive, while the other six drank only a placebo. The results of the experiment are given in Table 4.10.

Dr Fayle's colleague, Dr Suckseed, is also investigating the effect of alcohol on reaction times. He carries out an experiment using the simulated vehicle but he uses only 6 subjects compared with Dr Fayle's 12. However, each of the six subjects 'drives' the vehicle twice, once while under the influence of alcohol and once while sober. The results of this experiment are given in Table 4.11.

These two experiments are referred to in questions (1)–(10). You might care to check your answers to earlier questions before attempting later ones.

  (1) Which type of significance test would you carry out on Dr Fayle's data to answer the question 'Does alcohol cause an increase in reaction time?'

  (a) a one-sample t-test;
  (b) a two-sample t-test;
  (c) a paired comparison test;
  (d) none of the above.

  (2) Which type of significance test would you carry out on Dr Suckseed's data to answer the question 'Does alcohol cause an increase in reaction time?'

  (a) a one-sample t-test;
  (b) a two-sample t-test;
  (c) a paired comparison test;
  (d) none of the above.

(3) In the significance test chosen in question (2) what would be the standard deviation used when calculating the calculated value?

(a) 0.036 56
(b) 0.060 25
(c) 0.071 38
(d) 0.082 38

(4) In the significance test chosen in question (2) what would be the required value?

(a) 2.20
(b) 2.23
(c) 2.45
(d) 2.57

(5) In the significance test chosen in question (2) what conclusion would you draw?

(a) The evidence from Dr Suckseed's experiment is inconclusive.
(b) Reaction time is not affected by alcohol.
(c) Reaction time is decreased on average by the effect of alcohol.
(d) Reaction time is increased on average by the effect of alcohol.

(6) In the significance test chosen in question (1) what would be the calculated value?

(a) 1.29
(b) 1.42
(c) 1.78
(d) 2.83

(7) In the significance test chosen in question (1) what conclusion would you draw?

(a) The evidence from Dr Fayle's experiment is inconclusive.
(b) Reaction time is not affected by alcohol.
(c) Reaction time is decreased on average by the effect of alcohol.
(d) Reaction time is increased on average by the effect of alcohol.

(8) In what way is Dr Suckseed's experiment superior to Dr Fayle's experiment?

(a) By using fewer subjects Dr Suckseed has reduced the variability from person to person.
(b) We can calculate differences from Dr Suckseed's data, thereby eliminating the variation from drive to drive of each subject.
(c) The differences calculated from Dr Suckseed's data have a smaller standard deviation than the differences calculated from Dr Fayle's data.
(d) Dr Suckseed's data can be analysed by a paired comparison test which is more powerful than the two-sample t-test.

(9) In what way might Dr Suckseed's experiment be inferior to Dr Fayle's experiment?

    (a) A subject might improve with practice and this could affect Dr Suckseed's data but not Dr Fayle's.

    (b) A subject might be more fatigued during his or her second run which would make the comparison of Dr Fayle's sample invalid.

    (c) Some subjects might react more quickly after alcohol and some more slowly, which would invalidate the paired comparison test.

    (d) Some subjects might be more variable than others, which would invalidate the paired comparison test but not the two-sample t-test.

(10) Can we reasonably conclude, using all relevant data in Tables 4.10 and 4.11, that the mean reaction time of all sober drivers is less than 0.4 seconds?

Table 4.10 — Results of Dr Fayle's experiment

Person	A	B	C	D	E	F	G	H	I	J	K	L
Alcohol	Y	Y	N	Y	N	Y	Y	N	Y	N	N	N
Reaction time (seconds)	0.39	0.41	0.37	0.38	0.40	0.27	0.50	0.34	0.36	0.16	0.31	0.34

Table 4.11 — Results of Dr Suckseed's experiment

Person	U	V	W	X	Y	Z
Reaction time (seconds)						
Sober	0.31	0.34	0.34	0.16	0.37	0.40
After alcohol	0.39	0.41	0.38	0.27	0.50	0.36

    (a) Yes, because the mean reaction time for all 12 sober drivers is significantly less than 0.4 seconds.

    (b) Yes, because the majority of subjects had a reaction time less than 0.4 seconds during sober drives.

    (c) No, because the mean reaction time for all 12 sober drivers is not significantly less than 0.4 seconds.

(d) No, because two subjects did not have a reaction time less than 0.4 seconds during a sober drive.

4.12 ANSWERS TO SELF-TEST QUESTIONS

(1) (b)
(2) (c)
(3) (b)
(4) (d)
(5) (d)
(6) (b)
(7) (a)
(8) (d)
(9) (a)
(10) (c)

5

Safer significance testing

5.1 INTRODUCTION

In Chapter 4 you were asked to carry out several significance tests. I hope that you followed the recommended four-step procedure, in which a hypothesis is put forward and then a decision about the truth of this hypothesis is made in the light that is shed by the data. We shall continue to use the four-step procedure throughout this chapter, in which your repertoire of significance tests will be further extended. After you have completed this chapter you will, I hope, be convinced that carrying out a significance test is easy. However, while working through the chapter you may grow to realize that it is rather more difficult to select the most appropriate test and to ensure that the assumptions underlying the test are satisfied.

Fortunately these difficulties can be overcome provided that we exercise care in test selection and we check that the data are compatible with the associated assumptions. We will start by reconsidering the assumptions underlying the *t*-tests we carried out in Chapter 4. The aims of this chapter are as follows.

(1) The primary aim is to extend the repertoire of significance tests you built up while studying Chapter 4. The benefits for you will be twofold. Firstly, the additional significance tests will be of use in your analysis of data. Secondly, you will acquire a deeper understanding of the essentials of significance testing.

(2) The second aim of this chapter is to discuss in more detail the assumptions which underlie the *t*-tests in Chapter 4, and the confidence intervals in Chapters 2 and 3. I shall introduce alternative significance tests which you can use when you suspect that the assumptions are not satisfied.

(3) The discussion of assumptions would not be complete without some mention of outliers, flyers or rogues. These are of interest to all who wish to draw conclusions from experimental data. The third aim of this chapter, therefore, is to introduce a simple outlier test, which will help you to decide whether or not your more extreme observations come from the same population as the rest of the data.

5.2 ASSUMPTIONS UNDERLYING THE *t*-TEST

The **one-sample *t*-test** is used when we wish to decide whether or not a population mean is equal to some specified value. An alternative is to calculate confidence limits for the population mean and then to ask 'Does the interval include the specified value?' The *t*-test and the confidence interval will always lead us to the same conclusion, as we saw in Chapter 4. Thus it will not surprise you to learn that the two procedures share exactly the same assumptions. These are:

(a) The sample was selected at random from the population.

(b) The measured variable has a normal distribution.

We first discussed these assumptions in Chapter 1 and concluded that, in practice, they are unlikely to be satisfied entirely. However, we took consolation from the belief that the second assumption is no cause for concern unless the sample is very small and/or the population distribution is very skewed. However, sampling can be a problem in many situations, for no statistical analysis can protect us against biased sampling.

We used the **two-sample *t*-test** to compare two sample means in order to decide whether or not the two population means differed. An alternative procedure is to calculate confidence limits for the difference between the two population means and then to ask 'Does the interval include zero?' The two-sample *t*-test and the confidence interval will always lead us to the same conclusion. Thus it comes as no surprise, yet again, to learn that the two approaches share exactly the same assumptions. These assumptions are three in number:

(a) The two samples were selected at random from the two populations.

(b) The measured variable has a normal distribution in both populations.

(c) The two population distributions are equally variable.

The third assumption, concerning equal variability of the two populations, has not been mentioned explicitly in previous chapters. However, it was hinted at when we calculated confidence limits for the difference between two population means, in Chapter 2. To obtain the limits we used a combined standard deviation and I pointed out that it would be most unreasonable to combine two or more standard deviations if they differed considerably.

Perhaps this assumption of equal variability will be more meaningful if we consider a specific situation. Dr Blight wishes to assess the effectiveness of a liquid rose treatment. He has already demonstrated its ability to eliminate certain diseases but he is concerned that its use might inhibit growth. An experiment is set up in which 50 climbing rose shrubs are grown under normal conditions but 25 are treated and the other 25 left untreated. Several variables are measured on each shrub, including height, number of blooms, number of leaves, etc. The data on the number of blooms are illustrated in Fig. 5.1.

We can see in Fig. 5.1 that the number of blooms varies from bush to bush. However, we can also see that the number is **more variable** on the 25 untreated bushes than on those which were treated. Common sense tells us that it would be foolish to ask 'What effect does the treatment have on the **mean** number of blooms?' without also asking 'What effect does the treatment have on the **variability**?' The third of the three assumptions puts the point even more strongly. The assumption is telling us that it would **not be valid** to compare two sample means if the samples differed considerably in their variability.

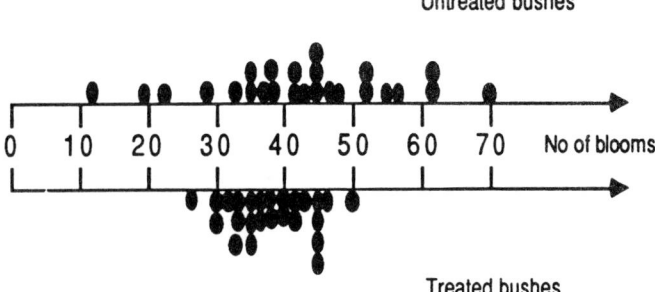

Fig. 5.1 — Comparison of treated and untreated bushes.

Suppose we had not examined Fig. 5.1 but plunged straight into a two-sample *t*-test. After calculating the necessary means and standard deviations, which are given in Table 5.1, we would see that one standard deviation was more than twice as large as the other. This would alert us to the danger, and it is wise to develop the habit of comparing sample standard deviations **before** comparing sample means.

Obviously, we are extremely unlikely to get two standard deviations which are exactly equal, whenever we take two samples. The sample standard deviation is going to vary from sample to sample, as the sample mean varies from sample to sample. Just how much must two standard deviations differ before we can confidently conclude that the two populations are not equally variable? The answer to this question is provided by a significance test known as the SD-test.

The purpose of an **SD-test** is to compare the variability of two populations. The null hypothesis of the SD-test is that 'the two populations are equally variable' and the calculated value is obtained from the sample standard deviations:

calculated value=(larger SD)/(smaller SD)

We shall now carry out an SD-test on the standard deviations in Table 5.1. If the calculated value is **greater** than the required value we will reject the null hypothesis and conclude that the two populations are **not** equally variable. Thus, if the calculated value were greater than the required value, we would be unwise to carry out a two-sample *t*-test, for the SD-test has shown that one of its assumptions is violated. If, however, the calculated value were less than the required value we would not reject the null hypothesis and we would then proceed with the *t*-test on the assumption that the two populations are equally variable.

SD-test

Null hypothesis   The two populations are equally variable.
Calculated value   =(larger SD)/(smaller SD)
                      =13.4166/5.8802
                      =2.28

Table 5.1 — Number of blooms on treated and untreated bushes

	Mean	Standard deviation
Treated	42.44	13.4166
Untreated	38.08	5.8802

Required value =1.50 (approximately) from Table ST4 with 24 and 24 degrees of freedom and 95% confidence.

Conclusion Because the calculated value is greater than the required value we reject the null hypothesis and conclude that the two populations are not equally variable.

The SD-test has told us that it is **not** valid to proceed with a two-sample *t*-test. Common sense and Fig. 5.1 have already told us that it would be unwise to restrict our analysis to a simple comparison of means. So the SD-test and common sense are not in conflict.

The two-sample *t*-test is a very powerful method of analysis with which to compare two means, but it is only of use when the two populations are similar in all respects except their means. Such a situation is represented in Fig. 5.2. However, we

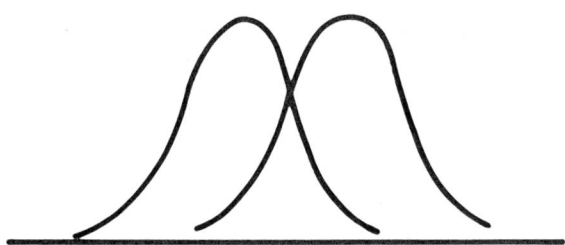

Fig. 5.2 — The *t*-test is valid.

should **not** use the two sample *t*-test in the type of situation shown in Fig. 5.3, where we have a clear difference in variability. The SD-test has suggested that our data came from the populations in Fig. 5.3 and not from the populations in Fig. 5.2.

Exercise 5.1
In the self-test of Chapter 2 you analysed a set of data which had resulted from an experiment in which two diets were compared. The final weights of the 15 pigs were as given in Table 5.2.

Table 5.2 — Weights of pigs from Chapter 2 self-test

Feed	Weight of pigs (kg)	Mean	SD
A	22.4, 26.6, 18.8, 20.2, 19.6, 24.5, 28.0, 19.9	22.50	3.48970
B	27.6, 30.0, 24.5, 30.2, 22.6, 28.7, 24.0	26.80	3.07734

(a) Draw a dot plot (like Fig. 5.1) to facilitate your comparison of the two samples.
(b) After examining your dot plot, are you prepared to believe that the two samples were drawn at random from two populations which had normal distributions and which were equally variable?
(c) Carry out an SD-test on the sample standard deviations.
(d) Carry out a two-sample t-test in order to compare the effectiveness of the two diets.

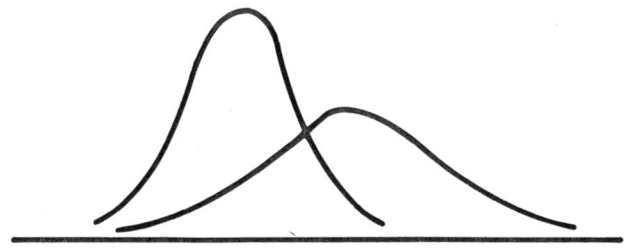

Fig. 5.3 — The t-test is not valid.

We have used the SD-test to check an assumption before doing a two-sample t-test. However, the SD-test is very useful in its own right. You will meet it again in Volume 3, *Statistics for Analytical Chemists*, when we will use it to compare the precision of two methods or two laboratories.

5.3 THE NORMAL DISTRIBUTION

In Chapters 1, 2, 3 and 4 I mentioned the assumptions underlying the statistical techniques as they were introduced. In each chapter the discussion of assumptions was rather brief, for I did not wish to bore you or to put you off using confidence intervals and/or significance tests. However, despite my brevity, you probably noticed that two assumptions recurred in every chapter. These were concerned with 'random sampling' and the 'normal distribution'. We shall now examine the normal distribution in more detail.

The four curves in Figs. 5.2 and 5.3 are known as normal distribution curves. They have a distinctive symmetrical bell shape. You could regard each curve as being the outline of a dot plot. Imagine that you were drawing a dot plot as you gathered more and more data. The shape of the dot plot would become smoother and approach the shape of the normal curve as your sample encompassed more and more of the population. Alternatively, you could regard the normal curve as the outline of a very smooth histogram, with bars so narrow that you could not see their width.

In practice we use a dot plot or a histogram to illustrate the distribution of a **sample**, whereas we use a normal curve to illustrate the distribution of an infinitely large **population**. Despite the mystery that surrounds anything infinite, normal curves follow some of the simple rules obeyed by dot plots and histograms. For example, the peak of a normal curve is above the mean of the population and the width of the curve depends on the standard deviation.

The two curves in Fig. 5.2 could represent the heights of adult males and females in the UK. Large-scale surveys have shown that the heights of fully grown males do appear to have a normal distribution, with a mean of 69 inches and a standard deviation of approximately 3 inches. The heights of females also appear to have a normal distribution, with the same standard deviation but a lower mean, 63 inches. Thus the two curves in Fig. 5.2 have the same width, but the female curve is 6 inches to the left of the male curve.

The two normal distribution curves in Fig. 5.3 could represent the heights of male children at two different ages. The older children are taller on average than the younger children. However, the older children are also more variable in height than the younger. Thus the normal curve for the older children is wider than the curve for the younger children, in addition to being further to the right.

The normal distribution is important for two reasons. The first is that, in practice, many sets of data look as if they could have come from a normal distribution. (The second reason is more complex and will be discussed later.) Whenever we believe that a sample has come from a population which has a normal distribution, we can make use of many statistical tables, including Tables ST1, ST2 and ST3 that we have used in earlier chapters. We can also make use of Table ST5 which tells us almost anything we might wish to know about any normal distribution.

Table ST5 actually describes what is known as the 'standard normal distribution'. This has the usual bell-shaped curve, with a mean equal to **zero** and a standard deviation equal to **one**. Obviously such a distribution is not likely to arise in any practical situation. Nonetheless, Table ST5 can be of use to us no matter what mean and standard deviation we have, provided that we first 'standardize' the value we wish to consider:

$$\text{standardized value} = (\text{value} - \text{mean})/\text{SD}$$

Consider, for example, the heights of adult males in the UK. What percentage of adult males are taller than 73 inches? I doubt whether you can answer this question from your observation of the people around you. However, if we are prepared to assume that the heights of adult males have a normal distribution with a mean height of 69 inches and a standard deviation of 3 inches, we can quickly obtain an answer from Table ST5. Before using the table it is wise to draw a diagram (Fig. 5.4).

The normal curve in Fig. 5.4 could be regarded as the outline of a dot plot which contained one dot for every adult male in the UK. We need to know what percentage

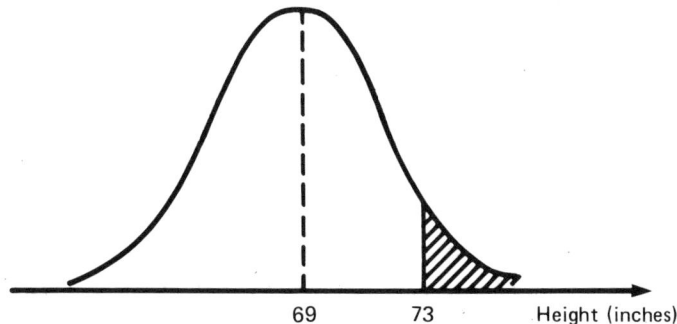

Fig. 5.4 — Males who are taller than 73 inches.

of the dots lie to the right of 73 inches. Thus we need to know what percentage of the area under the curve is in the shaded portion. First we must standardize the height in which we are interested (73 inches), then we can obtain the percentage from Table ST5.

standardized height = (height − mean)/SD
= (73 − 69)/3
= 1.33

For a standardized value of 1.33, Table ST5 gives a percentage equal to 9.2%. Thus we conclude that 9.2% of adult males in the UK are taller than 73 inches.

Exercise 5.2
Dr Brookfield has studied the production records for the triazone plant. The mean yield for the 230 batches produced in 1988 is 88.0 and the standard deviation is 7.8. A dot plot of the data indicates that the yield of triazone could well have a normal distribution.
(a) Use Table ST5 to estimate the percentage of future batches which will have yields greater than 100.
(b) Use Table ST5 to estimate the percentage of future batches which will have yield less than 80.
(c) What percentage of future batches will have yields between 80 and 100?

If you wish to carry out more calculations of the type in Exercise 5.2 you will have no difficulty finding such problems in conventional statistical texts. In my opinion certain topics are given excessive coverage in many texts, and the normal distribution is one of them. I would, therefore, like to return to more important issues, but before doing so I feel I should list some very useful information which applies to any normal distribution (Table 5.3). An awareness of the important features of normal distributions is useful when we attempt to check whether a sample came from a population which has a normal distribution.

When you are interpreting Table 5.3 you need to keep in mind that a standardized value tells you 'how many standard deviations from the mean is the value'. Thus a standardized value of $+1.7$ tells us that the value is 1.7 standard deviations above the mean. Similarly, a standardized value of -3.2 tells us that the value is 3.2 standard deviations below the mean. We see that the 95% in the right-hand column of Table 5.3 corresponds to a standardized value of 1.96. Thus we can conclude that 95% of the data from a normal distribution will lie within 1.96 standard deviations of the mean.

5.4 CHECKING FOR NORMALITY

If we reverse the above statement we obtain a useful rule to check whether or not a set of data comes from a normal distribution (the rule will be easier to remember if we round the 1.96 to 2.0).

During your preliminary inspection of a set of data calculate the limits:

(sample mean) ± 2 (sample standard deviation)

If much more than 2.5% of the data are beyond either limit, you should suspect that the population does not have a normal distribution.

Obviously it would be unwise to apply this rule too rigidly, especially with a small set of data. Suppose, for example, you have 20 observations in your data. The rule suggests that you should expect only half of one observation beyond either limit. Common sense suggests, however, that it would be unwise to blow the whistle unless at least two observations were outside one of the limits.

An alternative check of normality is to plot the data on special graph paper, known as 'normal probability paper'. The scale on this special graph paper is chosen so that the points would lie on a straight line if we plotted a whole population which had a normal distribution. If we plot a sample from a normal population we should find that the points do not deviate greatly from a straight line.

The normal probability plot in Fig. 5.5 is based on the yield data for 20 batches of digozo blue pigment given in Table 5.4. To produce this graph I first put the yield values into ascending order then plotted them against the expected percentages from Table ST6.

You can see in Fig. 5.5 that the points lie roughly on a straight line. There is certainly no clear indication that a smooth curve would fit the points significantly better, so we conclude that the variation in yield from batch to batch follows a normal distribution.

If, before seeing Fig. 5.5, you had never encountered a normal probability plot, you might wonder what such a plot would look like if the sample had been taken from a population which did **not** have a normal distribution. Clearly it is desirable to see several normal plots before you base an important decision on one. The following exercise will help to enlarge your experience.

Table 5.3 — Important features of the normal distribution

Standardized value	Percentage in one tail	Percentage in two tails	Percentage in the centre
1.28	10%	20%	80%
1.64	5%	10%	90%
1.96	2.5%	5%	95%
2.33	1%	2%	98%
2.58	0.5%	1%	99%
3.09	0.1%	0.2%	99.8%
3.29	0.05%	0.1%	99.9%

Exercise 5.3
(a) Use the sheet of normal probability paper on p. 000 to produce a normal plot of the impurity data in Table 5.4.
(b) Do you consider that a smooth curve would fit the points much better than a straight line?
(c) Draw a dot plot of the impurity data and comment on the two plots.
(d) Calculate the mean and the standard deviation of the impurity data in Table 5.4. Do most of the data lie within the interval defined by mean $\pm2(SD)$?

5.5 CHECKING FOR OUTLIERS

The dot plot that you drew in Exercise 5.3 gives a **strong** indication that the data came from a skewed distribution. I say 'strong' because the indication is not based solely on **one** observation. If we removed the highest impurity, 7.2, there would still be an indication of skewness. If we then removed the second highest impurity, 5.7, the skewness would still be apparent. I mention this point because an impression of skewness could be given by the presence of just one rogue observation among 19 others which came from a normal population.

Earlier we used the SD-test to compare two standard deviations, before calculating the combined SD for use in a two-sample *t*-test. We failed to prove that the sample standard deviations were significantly different and we were, therefore, happy to combine them. In practice, of course, you will sometimes find that the two standard deviations **are** significantly different, and you will not feel that it is safe to proceed with the *t*-test.

Regardless of the outcome of the SD-test you will surely wish to examine the two sets of data using a double dot plot such as Fig. 5.1. One purpose of this visual inspection would be to ascertain whether either:

(a) an apparent difference in variability was simply due to one rogue observation, or

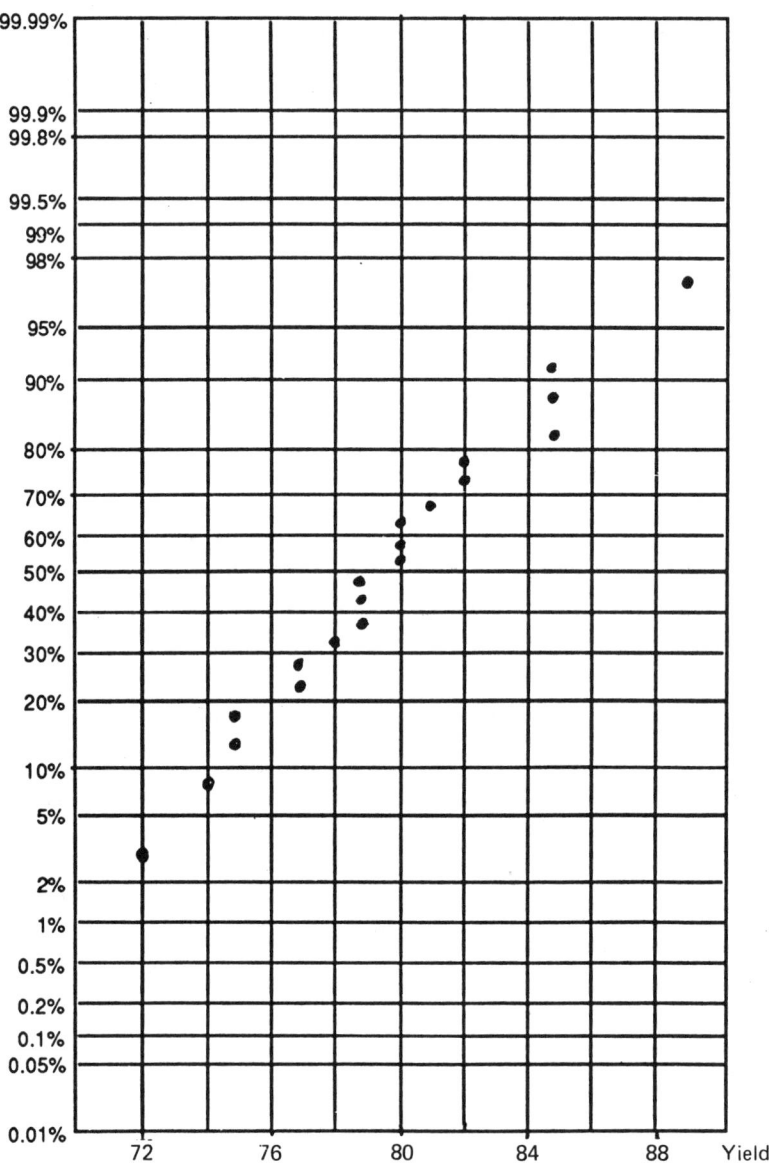

Fig. 5.5 — A normal probability plot.

(b) a real difference in variability was concealed by the presence of a rogue
    observation.

Rogues, or flyers, or outliers, as they are variously known, are of interest to all who
draw conclusions from data. Every scientist and technologist is aware that mishaps
can occur in any experiment. Obviously an unplanned occurrence can give rise to one

Table 5.4 — 20 batches of digozo blue pigment

Batch	A	B	C	D	E	F	G	H	I	J
Yield	82	79	85	77	80	74	82	75	72	78
Impurity	2.0	0.5	0.7	1.3	5.1	7.2	0.3	5.7	1.0	4.0

Batch	K	L	M	N	O	P	Q	R	S	T
Yield	81	79	89	75	80	80	77	85	85	79
Impurity	1.5	1.7	2.5	0.5	0.9	3.2	1.5	2.9	0.9	1.3

or more observations which do not fit into the pattern exhibited by the rest of the data. Thus, whenever a scientist finds that one or two observations do not appear compatible with the others, he or she may suspect that a mishap occurred, even though no incident was noted at the time. Of course, the scientist also realizes that data analysis becomes a nonsense if everyone feels free to discard data whenever it is convenient to do so. These are matters of widespread concern. I have been asked by many industrial clients, 'Is there an objective method for deciding whether or not an observation is an outlier?' and this question is invariably followed by, 'What should I do with an outlier once its existence has been established?'

The answer to the first question is 'yes, there are several methods, known as outlier tests'. Before we examine one such test, known as Dixon's test, let us have a definition.

A **suspected outlier** is an observation which does not appear to fit the pattern exhibited by the rest of the data. Usually the suspected outlier will be the largest or the smallest number in the data set. A **confirmed outlier** is an observation which has been rejected by an outlier test. .

Exercise 5.4
In order to determine the tensile strength of a batch of synthetic yarn five bobbins were selected at random from the 1000 bobbins in the whole batch, and then the breaking strain (kg) of a length of yarn from each of the five bobbins was measured. The results are:

28.7   26.5   27.4   23.4   27.9

(a) Draw a dot plot of the five determinations.
(b) I regard the lowest value, 23.4 kg, as a suspected outlier. Do you consider that the 23.4 is consistent with the pattern exhibited by the other four determinations?
(c) Calculate the three differences below and mark them on your dot plot:

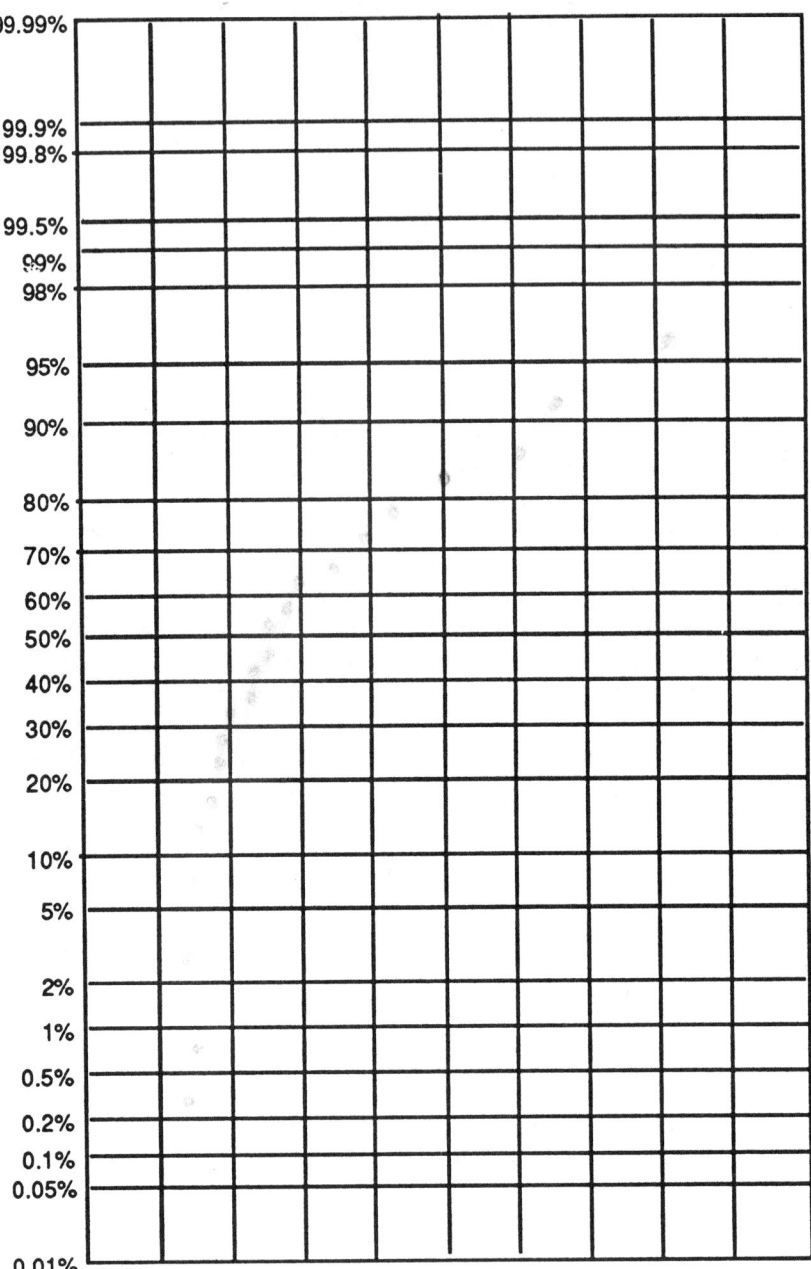

range=highest−lowest
lower gap=second lowest−lowest
upper gap=highest−second highest

(d) To carry out a Dixon's test you must now obtain the calculated value which is the larger of:

(lower gap)/(range)   or   (upper gap)/(range)

(e) Obtain the required value from the 95% column of Table ST7, for a sample size of five.
(f) What conclusions do you draw?
(g) (More difficult) Can you think of any reasons why Dixon's test should not be used with the impurity data in Table 5.4?

Outlier tests can be used to check on assumptions before carrying out other significance tests such as the *t*-test. However, it is important to realize that outlier tests, themselves, are based on assumptions. Obviously, an awareness of these assumptions is essential. If you felt a growing unease as you worked through Exercise 5.4, that may have been because you had no knowledge of the underlying assumptions of the test. Without such knowledge you could only guess the answer to Exercise 5.4(g), for example.

The two assumptions underlying Dixon's test are that the sample was selected at random from the population and that the population has a normal distribution. It is helpful to recognize this second assumption in the wording of the null hypothesis:

Dixon's test

Null hypothesis   All the data came from the same population which has a normal distribution.

Calculated value   is the greater of (upper gap)/(range), or (lower gap)/(range):
=0.587

Required value   =0.717 from the 95% column of Table ST7 for a sample size of 5.

Conclusion   As the calculated value is less than the required value we cannot reject the null hypothesis. Thus we are unable to conclude that the suspected outlier, 23.4 kg, did not come from the same normal distribution as the other four determinations.

When doing Exercise 5.4 you may have concluded that the lowest determination was an outlier. Dixon's test is telling us that the evidence is not sufficiently strong to warrant such a conclusion. In fact the lowest value would need to be as low as 20.9 kg to give a calculated value greater than the required value.

Now that I have revealed the assumptions underlying Dixon's test you will realize that it would be rather foolish to apply this test to the impurity data in Table 5.4. In our earlier examination the normal plot provided very strong evidence that the 20 impurity determinations came from a very skewed distribution. Thus it would not be very useful to use a less powerful method, Dixon's test, to prove that one of the impurity values did not come from the same normal distribution as the others.

Before we leave outlier testing I should like to point out that there are many other outlier tests, each of which offers certain advantages. In fact there are several 'Dixon's tests,' although the one we have used is certainly the most popular and is often referred to as **the** Dixon's test. For a more detailed coverage of outlier tests refer to Barnett and Lewis (1979).

5.6 A SAFER SIGNIFICANCE TEST

So far in this chapter I have reminded you of the assumptions underlying the t-tests and I have introduced several methods for checking the normal distribution assumption. Now it is time to discuss what action you might take if you suspect your population does not have a normal distribution. One possibility is to continue your data analysis using a significance test which has less demanding assumptions. To illustrate such a test I shall reconsider a set of data from an earlier chapter. In Chapter 4 we used a paired comparison test to check the effectiveness of a diet. Eight volunteers took part in the experiment and each was weighted before and after. The data are reproduced in Table 5.5.

Table 5.5 — A quantitative assessment of the diet

Person	A	B	C	D	E	F	G	H
Weight before	137	196	103	162	154	122	142	136
Weight after	130	190	98	153	156	115	143	133
Weight loss	7	6	5	9	−2	7	−1	3

In the paired comparison test we calculated the mean and standard deviation of the differences, and then we used these to obtain the calculated value. Because the calculated value exceeded the required value from Table ST1 we rejected the null hypothesis and concluded that the diet was effective. We then went on to calculate confidence limits which showed that the population mean weight loss was between 0.94 lb and 7.56 lb.

A scientist who had not enjoyed the benefits of a course in statistics might have adopted a different approach to the analysis of this data. Perhaps this less-numerate scientist would have focused attention on each person in turn and asked 'Did this person lose weight while on the diet?' The answers are given in Table 5.6.

We can see very clearly in Table 5.6 that six of the eight persons on the diet lost weight. This simple fact can be used as the basis of a significance test, known as the 'yes–no test'. Let us carry out this very simple test and then compare the conclusion with that drawn in Chapter 4 when we used the paired comparison test.

Table 5.6 — A qualitative assessment of the diet

Person	A	B	C	D	E	F	G	H
Did the person lose weight?	Yes	Yes	Yes	Yes	No	Yes	No	Yes

Yes–no test

Null hypothesis	50% of the population would answer 'yes' to the question 'Did you lose weight whilst on the diet?' (i.e. the diet has no effect).
Calculated value	=The number of 'yes's or the number of 'no's, whichever is greater: =6
Required value	=7.5 from Table ST8 with a sample size of 8 and 95% confidence.
Conclusion	Because the calculated value is less than the required value we cannot reject the null hypothesis. Thus we are unable to conclude that the diet is effective.

No doubt you are impressed by the simplicity of the yes–no test. When using this test we simply count how many times something occurred and compare this with a required value. What could be easier? However, you may also be rather disturbed by the yes–no test, because it has **not** led us to the same conclusions as that reached in the paired comparison test. When two alternative significance tests lead us to different conclusions which one are we to trust?

It can be shown that the paired comparison test is much more powerful than the yes–no test. This is because the paired comparison test makes full use of the data in Table 5.5, whereas the yes–no test is based on Table 5.6. When we condense quantitative data into qualitative data we achieve a greater simplicity but we also lose a great deal of information. The yes and no entries in Table 5.6 tell us which people lost weight but they do not tell us **how much** weight each person lost or gained. Thus there is much more information in Table 5.5 than in Table 5.6.

If the yes–no test is so wasteful of information why would any scientist wish to use it, in preference to the paired comparison test which is so much more powerful? One possible explanation is that the scientist would prefer the yes–no test because he or she is aware of the dangers inherent in the use of the paired comparison test. It is certainly true that the whole family of *t*-tests, which are much loved by physical scientists, are greatly distrusted by many social scientists. The basis of this distrust is the fear that the assumptions underlying the *t*-test will not be satisfied. In particular it is the normal distribution assumption that causes concern.

Significance tests can be grouped into two classes, which are often referred to as 'parametric tests' and 'non-parametric tests'. The difference between the two types of test is that the parametric tests are based on assumptions about the population distribution. The *t*-tests are obviously parametric tests as they share the assumption that the population has a normal distribution. In contrast, the yes–no test is a non-parmetric test. In general a non-parametric test will be less powerful than an equivalent parametric test. However, the non-parametric test will be safer because it is not dependent on assumptions about the population distribution.

There are many situations in which we do not have a choice between a parametric test and a non-parametric test. This will be the case if our raw data are essentially qualitative in nature. Consider, for example, preference testing of alternative products by a market researcher. Each participant in the experiment is presented with two products and asked to say which one he or she prefers. Thus the raw data are qualitative at the outset. There is no question of condensing quantitative data, like Table 5.5, into qualitative data, like Table 5.6. It will be clear from their answers which product is preferred by the participants; but these people are only a **sample** of potential customers. Which product would have been preferred if we had asked the whole population? The yes–no test can help us to decide.

Exercise 5.5

Dr Chewsey has been asked to carry out the consumer testing of two shampoos. Participants in this experiment are asked to use both shampoos and to decide which one they prefer. Of the 50 people who take part in the experiment 35 express a preference for brand X and only 15 for brand Y.

(a) Carry out a yes–no test to see whether Dr Chewsey can reasonably claim that one brand is preferred to the other.
(b) Can you think of any reason why Dr Chewsey's conclusions in part (a) might be invalid?

5.7 A RANKING TEST

In the previous section we carried out a yes–no test to check the effectiveness of a diet. Having analysed the same data in Chapter 4 using a paired comparison test, we were disturbed to find that the two tests did **not** lead us to the same conclusion. We noted that the difference could be explained by the fact that the yes–no test is less powerful than the paired comparison test. We also noted that the yes–no test is a non-parametric test and therefore is not dependent on the normal distribution assumption. To help you get these tests in perspective it should be pointed out that we were comparing two extremes of a continuum. The paired comparison test is the **most** powerful test that could be used with this data, whilst the yes–no test is the **least** powerful. Furthermore, there are other tests between the two extremes which might be more appropriate than either. One such test is the 'Wilcoxon matched pairs test' that we shall now explore.

Before we carried out the yes–no test we converted the differences in Table 5.5 into the qualitative data of Table 5.6. For the Wilcoxon test we shall not treat our data so brutally; we shall convert the differences into **rankings**. These rankings are given in Table 5.7. Certain rules must be observed as we convert the differences (i.e. the weight losses) into rankings. Firstly we ignore the sign of the differences.

Table 5.7 — Assessment of the diet by rankings

Person	A	B	C	D	E	F	G	H
Weight loss	7	6	5	9	−2	7	−1	3
Ranking	6.5	5	4	8	2	6.5	1	3

Secondly we must give a rank of 1 to the smallest difference. Thirdly we must ensure that people who have equal differences share the appropriate ranks equally. Thus persons A and F must share ranks 6 and 7, which gives each a rank of 6.5. Having obtained the rankings in Table 5.7 we must now calculate a rank total for those persons who lost weight and a rank total for those who gained weight. These rank totals are 33 (i.e. 6.5+5+4+8+6.5+3) and 3 (i.e. 2+1). Our calculated value is the larger of the two rank totals.

In a significance test based on ranks the null hypothesis refers to the population **median.** With a *t*-test, of course, the hypothesis refers to the population **mean.** Perhaps you realize that the median is more helpful than the mean when we are dealing with a skewed distribution. For example, it is usual to quote median salaries, because salaries within a particular profession are likely to have a skewed distribution. Thus ranking tests are preferred to *t*-tests when we suspect that the population distribution might be highly skewed.

Wilcoxon's matched pairs test

Null hypothesis   The population median difference is equal to zero (i.e. the diet has no effect).

Calculated value   =The greater of the rank total for the positive differences, or the rank total for the negative differences: =33

Required value   =32.5 from Table ST9 for a sample size of 8 and 95% confidence.

Conclusion   Because the calculated value is greater than the required value we reject the null hypothesis and conclude that the population median difference is not equal to zero. Thus we have proved beyond reasonable doubt that the diet is effective.

The Wilcoxon test has led us to the same conclusion as that reached when we carried out the paired comparison test. Thus the ranking test and the *t*-test are in agreement with each other and both are at odds with the yes–no test. It can be shown that the Wilcoxon matched pairs test is much more powerful than the yes–no test, but not so powerful as the *t*-test.

In the light of what has been said about these three tests, which is the best one to use for assessing the effectiveness of the diet? It is certainly not wise to use the yes–no test. Although this simple test is perfectly valid, it is wasteful of information. Thus we are left with a choice between the *t*-test and the ranking test. If the assumptions underlying the paired comparison test are likely to be satisfied then we should use it. If, however, we suspect that the population of differences does not have a normal distribution we should use the ranking test.

In some situations it is difficult to decide whether it would be wise to use a parametric or a non-parametric test. In other situations there is no doubt. If, for example, we wish to draw conclusions from the data in Exercise 5.5, we **must** use the yes–no test. It is not possible to use the *t*-test or the ranking test because the raw data are qualitative in nature and therefore **cannot** be converted into rankings or into differences.

Exercise 5.6

Dr Chewsey, who carried out the consumer testing of two shampoos in Exercise 5.5, wishes to compare two hair sprays. He could repeat his earlier experiment, with a large number of participants being asked to try both products and express a preference for one or the other. The results would be analysed using a yes–no test.

However, Dr Chewsey wishes to make use of a more powerful significance test so that he will not need such a large sample. He therefore resolves to use Wilcoxon's matched pairs test. In his experiment he asks each of ten participants to give a score to each of the hair sprays. The results are given in Table 5.8.

Table 5.8 — Scores for two hair sprays

Person	A	B	C	D	E	F	G	H	I	J
Score for Brand P	2	1	3	1	3	1	4	2	1	1
Score for Brand Q	5	3	1	5	5	4	3	5	5	4

5=very good, 4=good, 3=average, 2=poor, 1=very poor.

(a) Name three significance tests which could be used to decide whether or not there was an overall preference for either brand.
(b) What would you take into account when deciding which of the three tests you would use?
(c) Carry out a Wilcoxon's matched pairs test to see whether brand Q is more favourably regarded by potential customers than brand P.

5.8 A SUMMARY OF THE IMPORTANT POINTS

(1) At the start of this chapter I reminded you of the assumptions underlying the *t*-tests that you used in Chapter 4.
(2) The two-sample *t*-test has the additional assumption that the two populations

Table 5.9 — Significance tests in chapters 4 and 5

Chapter	Type of test	Null hypothesis	Calculated value	Required value
4	Correlation test	The population correlation coefficient is equal zero	The sample correlation coefficient	From Table ST3: the value depends only on the sample size
4	One-sample t-test	The population mean is equal to some specified value	$[|\bar{x}-\mu|\sqrt{n}]/s$	From Table ST1 with degrees of freedom appropriate to the standard
4	Two-sample t-test	The two population means are equal	See below	From Table ST1, with degrees of freedom appropriate to the standard deviation
4	Paired comparison test	The population mean difference is equal to zero	$[|\bar{x}-\mu|\sqrt{n}]/s$	From Table ST1, with $n-1$ degrees of freedom
5	SD-test	The two populations are equally variable	(Larger SD/smaller SD)	From Table ST4, with degrees of appropriate to the standard deviations
5	Dixon's test	All the data came from the same population	Larger of A or B (see Table ST7)	From Table ST7: the value depends only on the sample size
5	Yes–no test	50% of the population would answer 'Yes' to the question	The number of 'yes's or the number of 'no's, whichever is the greater	From Table ST8: the value depends only on the sample size
5	Wilcoxon's matched pairs test	The population median difference is equal to zero	The greater of the two rank totals	From Table ST9: the value depends only on the sample size

In the two-sample t-test the calculated value $=(\bar{x}_1-\bar{x}_2)/\sqrt{\{s[(1/n_1)+(1/n_2)]\}}$

are equally variable. This assumption can be checked by carrying out an SD-test.

(3) All t-tests, and many other techniques, share the assumption that the population has a normal distribution. One characteristic of normal distributions is that 95% of observations will lie within two standard deviations of the mean. Thus whenever we find much of our data beyond two standard deviations from the the sample mean we must suspect that the population does not have a normal distribution.

(4) A useful way to assess the shape of the population distribution is to plot the data on normal probability paper. Curvature in such a plot is indicative of skewness.

(5) Excessive variability in a set of data may be indicative of a highly variable population, or it may simply result from the presence of an outlier.

(6) An outlier is an observation which does not belong to the same population as the bulk of the data. Suspected outliers can be checked by the use of an outlier test such as Dixon's test.

(7) Outlier tests themselves are based on assumptions and should be used with care. The rejection of data should be based on scientific reasoning wherever possible.

(8) Two non-parametric tests were introduced in this chapter. The yes–no test can be used with qualitative data and the Wilcoxon matched pairs test with ranked data.

(9) The advantage of non-parametric tests is that they are not dependent on assumptions about the population distribution. Thus they are safer to use than parametric tests, such as the t-tests. The disadvantage of non-parametric tests is that they are less powerful.

(10) All the significance tests introduced in Chapters 4 and 5 are summarized in Table 5.9.

5.9   ADDITIONAL EXERCISES

Exercise 5.7
Table 5.10 contains the filled weights of 20 bags of granulated fertilizer. They were 20 consecutive bags filled by one particular operator using a new bagging jig.

Table 5.10 — Weight (kg) of 20 bags of fertilizer

49.8	51.0	53.3	51.6	48.2	51.0	52.2	52.5	51.6	50.5
52.5	54.0	53.0	53.8	54.5	52.5	51.9	53.6	53.8	53.2

(a) Plot the weights on normal probability paper using expected population percentages from Table ST6.
(b) Draw a dot plot of the weights.
(c) Carry out a Dixon's test to see whether either the highest or lowest weights could reasonably be classed as outliers
(d) Are you happy to conclude that the data in Table 5.10 came from a population which has a normal distribution?

Exercise 5.8
It appears that the filling of the 20 bags in Exercise 5.7 was not carried out under homogeneous conditions. After filling the first 10 bags, which gave the weights in the top row of Table 5.10, the operator removed a protective shield, because it prevented the smooth operation of the flow lever. Naturally, you wish to know whether this modification had any effect on the weights of the filled bags.
(a) Draw two dot plots which will help you to compare the weights of the first ten bags with the weights of the second ten.
(b) Plot the two sets of data on the same sheet of normal probability paper that you used in Exercise 5.7. Keep the two sets quite separate and use expected population percentages for a sample size of 10.
(c) Carry out an SD-test to see whether one run of 10 batches is significantly more variable than the other.
(d) Do you consider that it would be reasonable to carry out a two-sample t-test in

order to ascertain whether or not this operator filled to a significantly higher weight on average after the shield was removed?

(e) Would it be appropriate to carry out a Wilcoxon matched pairs test in order to check whether or not this operator filled to a significantly higher weight on average after the shield was removed?

(f) Regardless of how you answered parts (d) and (e), carry out a two-sample *t*-test.

5.10 WORKED SOLUTIONS

Solution to Exercise 5.1

(a)

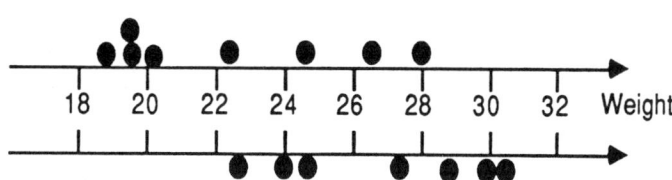

Fig. 5.6 — A double dot plot to compare two feeds.

(b) Both sets of data could well have come from normal distributions. There is no one point on the dot plot which is far removed from its nearest neighbour and, therefore, no indication of skewness. The data for feed A are more widely spread than those for feed B but the difference in variability is not striking. In my opinion these data do not contradict the assertion that the two populations have normal distributions which are equally variable.

(c) *Null hypothesis* The two populations are equally variable (i.e. pigs fed on diet A vary in weight just as much as pigs fed on diet B).

Calculated value =(Larger SD)/(smaller SD)
=3.489 70/3.077 34
=1.13

Required value =2.39 from Table ST4 with 7 and 6 degrees of freedom and 95% confidence.

Conclusion Because the test value is less than the table value we cannot reject the null hypothesis. Thus we are unable to conclude that the populations are not equally variable.

(d) In the SD-test in part (c) we were unable to demonstrate that the two populations differed in variability. We shall therefore, assume that the two populations are equally variable, which is necessary if we are to proceed with a two sample *t*-test.

Null hypothesis The two populations have equal means (i.e. the two diets are equally effective).

Calculated value $=(\bar{x}_1-\bar{x}_2)/\sqrt{\{s[(1/n_1)+(1/n_2)]\}}$
$=(26.80-22.50)/\sqrt{\{3.3058[(1/7)+(1/8)]\}}$
$=2.51$

Required value	=2.16 from Table ST1 with 13 degrees of freedom and 95% confidence.
Conclusion	Because the test value is greater than the table value we reject the null hypothesis and conclude that the two diets are not equally effective. Feed B with a mean of 26.8 kg gives a greater weight increase than does feed A with a mean of 22.5 kg.

Solution to Exercise 5.2

(a) Standardized yield=(yield−mean)/SD
$$=(100-88)/7.8$$
$$=1.54$$

With a standardized value of 1.54 we obtain a percentage of 6.2% from Table ST5. Thus we conclude that 6.2% of future batches will have yields greater than 100.

(b) Standardized yield=(yield−mean)/SD
$$=(80-88)/7.8$$
$$=-1.03$$

The standardized value is negative because the yield is less than the mean. Unfortunately there are no negative standardized values in Table ST5. Because the normal distribution is symmetrical it is not necessary to tabulate both halves. In practice, therefore, whenever the standardized value is negative we ignore the minus sign. With a standardized value of 1.03 we obtain a percentage of 15.2% from Table ST5. Thus we conclude that 15.2% of future batches will have yields less than 80.

(c) The percentage of future batches with yield between 80 and 100 can be calculated from the percentages you obtained in parts (a) and (b):

the required percentage=100−(6.2+15.2)
$$=78.6\%$$

We conclude that 78.6% of future batches will have yields between 80 and 100. If you had any difficulty with this question you may find the diagram in Fig. 5.7 helpful.

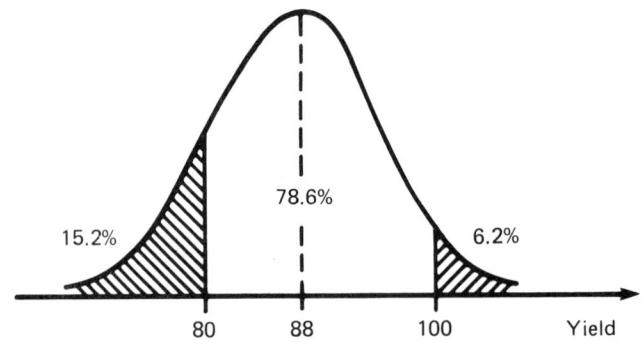

Fig. 5.7 — Illustration of solution to Exercise 5.2(c).

Solution to Exercise 5.3

(a)

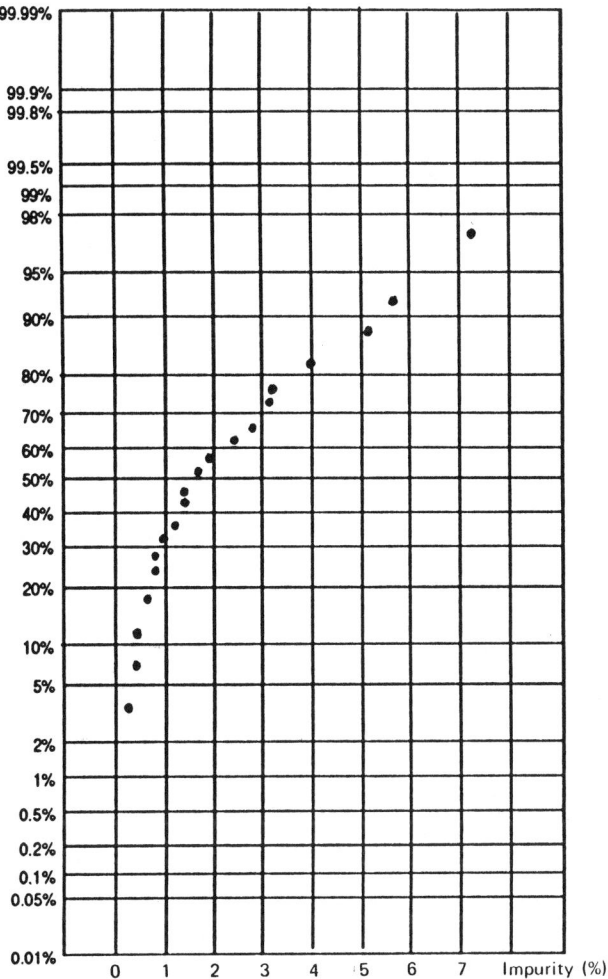

Fig. 5.8 — Normal plot of the impurity data.

(b) In my opinion a smooth curve would fit the points much better than a straight line.

(c)

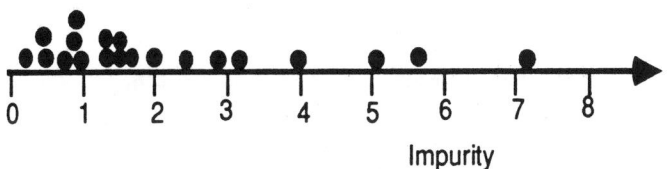

Impurity

Fig. 5.9 — A dot plot of the impurity data.

We can see in the dot plot (Fig. 5.9) that the batch-to-batch variation in impurity has a very skewed distribution. It is this skewness which gives rise to the curve in the normal probability plot.

(d) Mean=2.235, SD=1.9168. Therefore

$$\text{mean}\pm2(\text{SD})=2.235\pm2(1.9168)$$
$$=2.235\pm3.834$$
$$=-1.599 \text{ to } 6.069$$

We see that one of the twenty impurity determinations lies above the upper limit while the lowest impurity is well above the lower limit. Thus there is an indication of skewness. Clearly the normal plot gives a much clearer indication of skewness.

Solution to Exercise 5.4

(a)

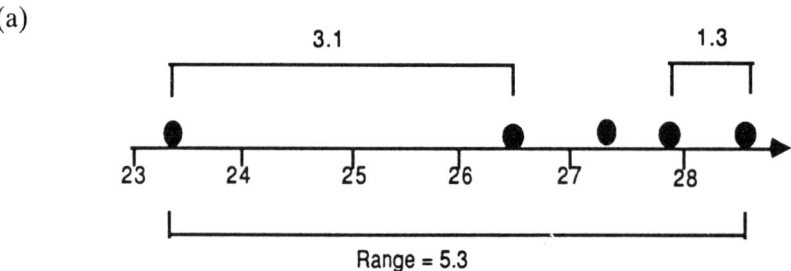

Fig. 5.10 — Solution to Exercise 5.4(a).

(b) The gap between the lowest value, 23.4, and the second lowest, 26.5, does seem rather large. Perhaps, with such a small sample, a relatively large gap is quite likely to occur. I suspect you may have difficulty reaching a decision. Furthermore, while pondering the problem you may conclude that you do not understand the question.

(c) Range=28.7−23.4=5.3 kg; lower gap=26.5−23.4=3.1 kg; upper gap=28.7−27.9=0.8 kg.

(d) Calculated value=0.585.

(e) Required value=0.717.

(f) As the calculated value is less than the required value we cannot reject the null hypothesis. However, as I have not put forward a null hypothesis you may not know what conclusion to draw. We shall discuss this point when you return to the text.

(g) This is a difficult question for anyone who has not used Dixon's test before. A very astute reader may have realized that Dixon's test cannot be applicable to all data; there must be some assumption about the population distribution built into the table of required values. We shall discuss this point when you return to the text.

Solution to Exercise 5.5

(a) *Null hypothesis*   50% of the population would answer 'yes' and 50% would answer 'no' (the question being, 'do you prefer brand X?')

     Calculated value   =The number of 'yes's or the number of 'no's, whichever is greater:

           =35

     Required value   =32.5 from Table ST8 with a sample size of 50 and 95% confidence.

     Conclusion   Because the calculated value is greater than the required value we reject the null hypothesis. We conclude that brand X is preferred to brand Y.

(b) Dr Chewsey does not need to worry about the normal distribution when he is using the yes–no test. However, there is one assumption which cannot be avoided whatever significance test is used. This is the random sampling assumption. If Dr Chewsey's 50 participants are not a **random sample** of potential customers for these shampoos then he should give serious thought to the value of his conclusion. Furthermore, if the 50 participants are obviously **not representative** of all potential customers Dr Chewsey could well be misled by the yes–no test.

Solution to Exercise 5.6

(a) Three significance tests which could be used to analyse these data are:
    (i) the paired comparison test which was introduced in Chapter 4;
    (ii) the Wilcoxon test;
    (iii) the yes–no test.

(b) While considering which of the three tests is most appropriate we need to consider the **power** of each test and the **assumptions** underlying each test. The paired comparison test is the most powerful of the three but it has the most demanding assumptions. The yes–no test and the Wilcoxon test are less powerful but they do not require that the differences should have a normal distribution. In fact the yes–no test can be used in situations where we do not have any differences.

    If you calculate differences from the pairs of scores you may feel that the normal distribution assumption is unlikely to be satisfied. In that case the Wilcoxon test would be the best significance test to use.

(c)

Table 5.11 — Data for Exercise 5.6(c)

Person	A	B	C	D	E	F	G	H	I	J
Difference	3	2	−2	4	2	3	−1	3	4	3
Ranking	6.5	3	3	9.5	3	6.5	1	6.5	9.5	6.5

The rank total for those persons who prefer brand P=4; rank total for those persons who prefer brand Q=51.

Null hypothesis   The population mean difference is equal to zero (i.e. the two
                       brands are equally liked).
Calculated value   =51
Required value   =46.5 from Table ST9 for a sample size of 10.
Conclusion   Because the calculated value is greater than the required
                       value we reject the null hypothesis and conclude that the
                       population mean difference is not equal to zero. Thus we
                       conclude that brand Q is preferred to brand P.

Solution to Exercise 5.7
(a)

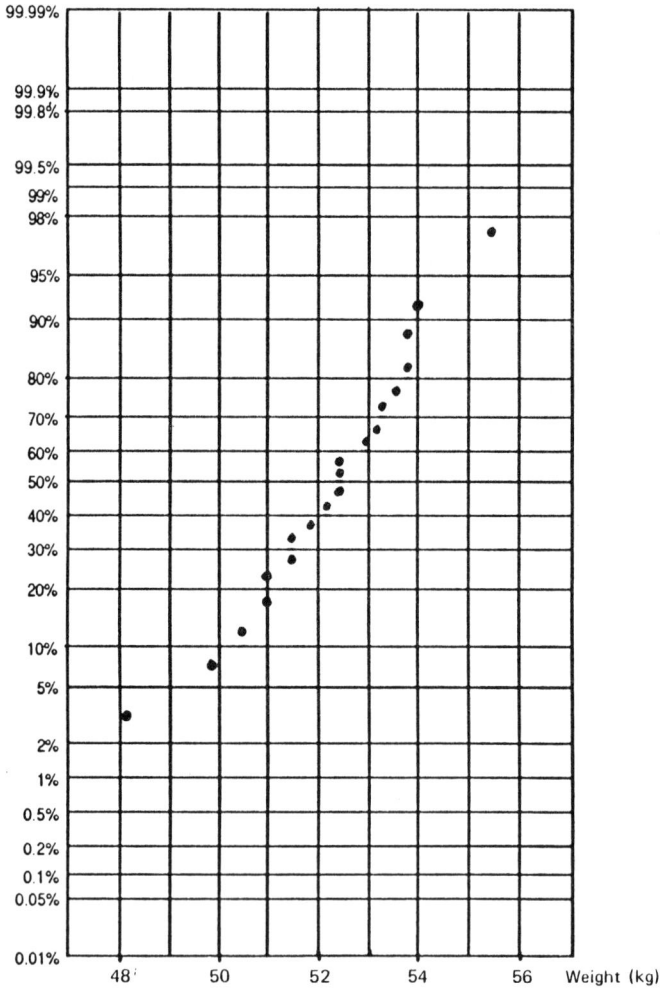

Fig. 5.11 — Weights of 20 bags of fertilizer.

(b)

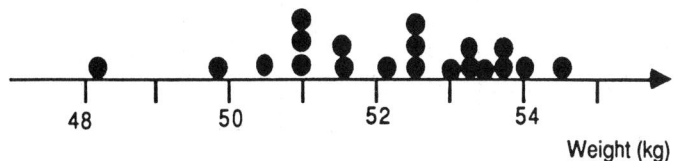

Fig. 5.12 — Weights of 20 bags of fertilizer.

(c) *Null hypothesis*   The 20 weights came from the same population which has a
                           normal distribution.
    Calculated value   =The greater of (49.8−48.2)/54.5−48.2) or
                           (54.5−54.0)/(54.5−48.2):

                           =0.254
    Required value   =0.34 from Table ST7 for a sample size of 20.
    Conclusion   As the calculated value is less than the required value we
                           cannot reject the null hypothesis. Thus we cannot reason-
                           ably conclude that the lowest weight, 48.2 kg, is an outlier.

(d) There is some evidence of skewness in the dot plot and in the normal plot.
    However, this evidence is not strong and, with such a large sample, one would be
    happy to regard the 20 weights as having been sampled from a normal
    distribution.

Solution to Exercise 5.8
(a)

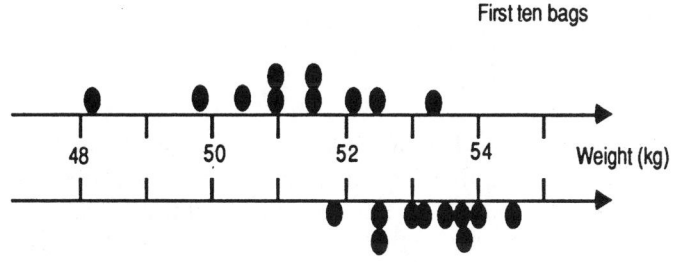

Fig. 5.13 — Weights of 20 bags of fertilizer.

(b)

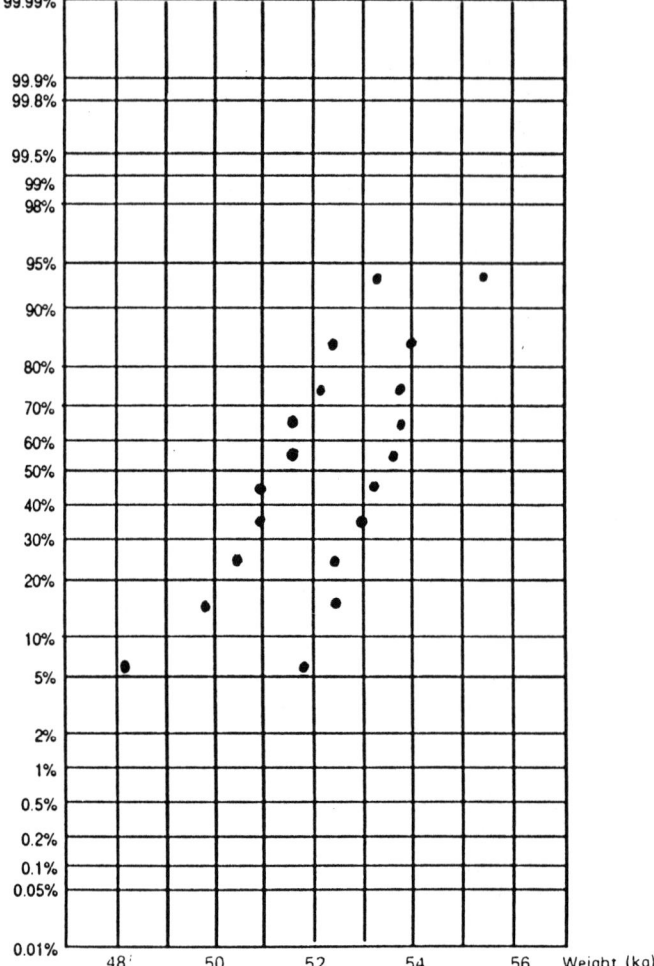

Fig. 5.14 — Weights of 20 bags of fertilizer.

(c) *Null hypothesis*   The two populations are equally variable (bag-to-bag variation in weight is the same, regardless of whether or not the shield is fitted).

     Calculated value   =(Larger SD)/(smaller SD)
                            =1.45071/0.80664
                            =1.80

     Required value   =2.01 from Table ST4, with 9 and 9 degrees of freedom.
     Conclusion   As the calculated value is less than the required value we cannot reject the null hypothesis. Thus we are unable to conclude that removing the shield affects the consistency of the bagging process.

(d) There is some evidence that one population is more variable than the other. Furthermore, there is some evidence that the greater variability is due to an outlier. However, we have shown that this evidence is not conclusive. I would be happy to carry out a two-sample t-test.

(e) The Wilcoxon matched pairs test is an alternative to the paired comparison t-test. It is **not** an alternative to the two-sample t-test. Thus it would **not** be appropriate to use a Wilcoxon matched pairs test in this situation. In the next chapter we shall consider a second Wilcoxon test, known as the 'signed ranks test', which is an alternative to the two-sample t-test.

(f) *Null hypothesis*   The two population means are equal (mean weight is the same regardless or whether or not the shield is fitted).

Calculated value $= (\bar{x}_1 - \bar{x}_2)/\{s\sqrt{[(1/n_1)+(1/n_2)]}\}$
$= (53.28 - 51.17)/\{1.1737\sqrt{[(1/10)+(1/10)]}\}$
$= 4.02$

Required value $= 2.10$ from Table ST1 with 18 degrees of freedom.

Conclusion   As the calculated value is greater than the required value we reject the null hypothesis and conclude that the mean weight did change significantly when the shield was removed.

5.11   DETAILED OBJECTIVES FOR THIS CHAPTER

If you have studied this chapter carefully and reflected on how the new concepts relate to your previous knowledge, you should be able to do the following.

(1) Explain the meaning of the following terms and use them correctly in appropriate contexts:

    (a) normal distribution;
    (b) normal probability plot;
    (c) standardized value;
    (d) SD-test;
    (e) outlier;
    (f) suspected and confirmed outliers;
    (g) Dixon's test;
    (h) yes–no test;
    (i) Wilcoxon's matched pairs test;
    (j) non-parametric tests.

(2) Describe the essential features of the **normal distribution** and use Table ST5 to estimate appropriate percentages.

(3) Plot a set of data on **normal probability paper**, using expected population percentages from Table ST6, and draw conclusions about the shape of the population distribution.

(4) Use the **SD-test** to compare the variabilities of two sets of data.

(5) Use **Dixon's test** to check for outliers in a set of data.

(6) Explain the usefulness of the SD-test, Dixon's test and graphical techniques in the checking of **assumptions** underlying t-tests.

(7) Carry out a **Wilcoxon's matched pairs test** to compare two population medians.

(8) Carry out a **yes–no** test in appropriate circumstances.

(9) Explain the advantages and disadvantages of **non-parametric tests** compared with the parametric tests studied in Chapter 4.

5.12 SELF-TEST

Dr Snow is investigating the effect of high frequency sound on the development of young rats. He takes two rats of equal weight from each of seven litters. One rat from each pair is reared in a 'quiet' environment while the other is reared under 'noisy' conditions. After four weeks of this treatment the weights in grams of the 14 rats are as given in Table 5.12. This information is referred to in questions (1)–(6).

Table 5.12 — Weights of rats after four weeks

Litter	A	A	B	B	C	C	D
Rat	1	2	3	4	5	6	7
Treatment	N	Q	N	Q	N	Q	N
Weight (g)	126	137	155	153	116	119	165

Litter	D	E	E	F	F	G	G
Rat	8	9	10	11	12	13	14
Treatment	Q	N	Q	N	Q	N	Q
Weight (g)	171	132	139	121	125	136	134

(1) Which of the following significance tests would you use with the data in Table 5.12 to answer the question 'Is the growth of rats retarded by rearing in a noisy environment?'
   (a) a yes–no test;
   (b) a Wilcoxon matched pairs test;
   (c) an SD-test;
   (d) a two-sample t-test.
(2) Suppose Dr Snow had used one rat from each of 14 litters rather than two rats from each of seven litters. How would his data analysis have been affected by this change?
   (a) The same significance test would be used but it would not be based on differences.
   (b) The same significance test would be used but the assumptions would be less likely to be satisfied.
   (c) A different significance test would be appropriate.
   (d) The data analysis would **not** be affected by this change in the experiment.
(3) If you carried out a yes–no test on the data in Table 5.12, what would be the calculated value?
   (a) 2
   (b) 4

(c) 5

(d) 6

(4) If you carried out a Wilcoxon's matched pairs test on the data in Table 5.12 what would be the calculated value?

(a) 3

(b) 13

(c) 15

(d) 25

(5) If you intended to carry out a paired comparison t-test on the weights in Table 5.12, which of the following would help you to check the assumptions underlying the test?

(a) a Dixon's test on the seven differences;

(b) a normal probability plot of the 14 weights;

(c) an SD-test to compare the SD of the noisy weights with the SD of the quiet weights;

(d) a two-sample t-test to compare the noisy mean with the quiet mean.

(6) If you carried out a Dixon's test on the seven differences from Table 5.12 what would be the calculated value?

(a) 0.000

(b) 0.308

(c) 0.385

(d) 0.444

(7) Dr Snow knows from previous research that the weights of rats of this age would have a normal distribution with a mean of 150 g and a standard deviation of 10 g. What percentage of such rats would have weights between 135 g and 155 g?

(a) 24.2%

(b) 25.8%

(c) 62.4%

(d) 75.8%

(8) If a suspected outlier is confirmed by Dixon's test, then which of the following is correct?

(a) The confirmed outlier must be removed before a significance test is carried out.

(b) It is possible that the confirmed outlier came from the same population as the rest of the data.

(c) A non-parametric test must be used to analyse the data.

(d) None of the above.

(9) What is the main advantage of using a parametric test rather than a non-parametric test?

(a) The assumptions underlying the parametric test are more likely to be satisfied.

(b) If we use a parametric test the null hypothesis cannot be rejected.

(c) If we use a parametric test we do not need a random sample.

(d) A parametric test is more powerful.

(10) What is the main advantage of using a non-parametric test rather than a parametric test?

(a) A non-parametric test is not based on the assumption that the sample was drawn at random from the population.
(b) If we use a non-parametric test the null hypothesis cannot be rejected.
(c) If we use a non-parametric test we do not need to use the 95% confidence level.
(d) A non-parametric test is not based on any assumptions about the population distribution.

5.13 ANSWERS TO SELF-TEST QUESTIONS

(1) (b)
(2) (c)
(3) (c)
(4) (d)
(5) (a)
(6) (b)
(7) (c)
(8) (b)
(9) (d)
(10) (d)

6

More significance tests and confidence intervals

6.1 INTRODUCTION

I suspect that you may have mixed feelings about Chapters 4 and 5. On the one hand, you may be very pleased with the progress you have made, having mastered eight types of significance test. However, on the other hand, you may have a growing suspicion that information is being withheld from you.

It is true that I have attempted to present significance testing in the simplest possible way, so that you could learn how to **do** significance tests without being distracted by important but peripheral ideas. In this chapter I shall address these outstanding issues and in doing so I may well put your mind at ease. For example, I shall attempt to answer a question which you have probably already asked, 'Why do we always use the 95% confidence level and what would be the effect of using other levels?'

An important aim of this chapter is to tie the loose ends that may be dangling in your mind after studying Chapters 4 and 5. I shall attempt to expose the ways by which the incompetent and the unscrupulous may deceive others by their use of significance tests. These points may have been worrying you.

A second aim of this chapter is to extend your study of significance testing and confidence intervals into the less familiar world of qualitative data. The techniques we use with **qualitative** data are similar in many ways to those we have used with **quantitative** data in Chapters 1–5. However, there is one important difference in that much larger sample sizes are needed.

A third aim is to add to your repertoire two additional significance tests which are related to those introduced in Chapters 4 and 5. One of these new tests is a remarkably simple alternative to the two-sample t-test. Its very simplicity may help to reveal the essence of significance testing and thus consolidate your overall appreciation of this technique.

6.2 ANOTHER RANKING TEST

'When will it all end?', you may ask, 'Is there no limit to the number of significance tests a scientist needs to know?' In Chapter 4 we examined three types of t-test. Each is very powerful and very useful, but it was suggested in Chapter 5 that many researchers prefer to use alternative tests. Must you now study a host of non-parametric tests? No. We shall consider just three more tests before we close the list and declare you to be a fully qualified significance tester.

Let us now examine a non-parametric test which is an alternative to the two-sample t-test introduced in Chapter 4. You may recall that we used the two-sample t-test to compare two varieties of wheat. Each variety was planted in several plots of land and the yield in tons per hectare was measured for each of the plots. The data are reproduced in Table 6.1.

Table 6.1 — Yield of wheat from ten plots

Plot	1	2	3	4	5	6	7	8	9	10
Variety	A	A	B	A	B	B	A	A	B	A
Yield	3.9	4.6	3.0	4.1	3.5	3.3	3.9	3.4	2.2	2.6

Perhaps it would be useful to summarize the analysis carried out in Chapter 4. The means for the two varieties were 3.75 and 3.00 tons per hectare. The standard deviations were 0.683 37 and 0.571 55 which have a combined standard deviation equal to 0.643 72. The calculated value (1.80) was less than the required value (2.31) and we were, therefore, unable to conclude that either variety of wheat was superior to the other.

This two-sample t-test is only valid, of course, if the two populations have normal distributions. Furthermore, the two populations should be equally variable, as we noted in Chapter 5. However, the non-parametric test that can be used as an alternative to the t-test is **not** based on such assumptions. It is known as Wilcoxon's rank sum test (not to be confused with the Wilcoxon matched pairs test that we used in Chapter 5 when analysing paired data).

The first step to be taken in the rank sum test is to **rank** (Table 6.2) the yields in

Table 6.2 — Ranking of yields from ten plots

Plot	1	2	3	4	5	6	7	8	9	10	Total
Variety	A	A	B	A	B	B	A	A	B	A	
Yield	3.9	4.6	3.0	4.1	3.5	3.3	3.9	3.4	2.2	2.6	
Rank											
A	3.5	1		2			3.5	6		9	25
B			8		5	7			10		30

Table 6.1. There are, of course, two ways in which the ranks can be allocated. We could give a rank of 1 to the smallest yield or we could give a rank of 1 to the largest yield. We shall do the latter. The reason for this choice will be explained later.

The calculated value for the Wilcoxon rank sum test is the greater of the two rank totals. The six plots planted with variety A have a rank total of 25, whilst the four plots planted with variety B have a rank total of 30. Thus the calculated value is equal to 30.

Wilcoxon's rank sum test

Null hypothesis	The two population medians are equal (i.e. the two varieties of wheat give the same median yield).
Calculated value	=Larger of the two rank sums =30
Required value	=31.5 from Table ST10 with sample sizes of 4 and 6 and 95% confidence.
Conclusion	Because the calculated value is less than the required value we cannot reject the null hypothesis. Thus we are unable to conclude that either variety of wheat is superior to the other.

To obtain the ranks in Table 6.2 we gave a rank of 1 to the plot with the largest yield and a rank of 10 to the plot with smallest yield. We could have obtained equally valid rankings if we had used the opposite convention, giving a rank of 1 to the smallest yield. These alternative rankings are given in Table 6.3.

Table 6.3 — Unsuitable rankings of yield

Plot	1	2	3	4	5	6	7	8	9	10	Total
Variety	A	A	B	A	B	B	A	A	B	A	
Yield	3.9	4.6	3.0	4.1	3.5	3.3	3.9	3.4	2.2	2.6	
Rank											
A	7.5	10		9			7.5	5		2	41
B			3		6	4		1			14

The alternative rankings in Table 6.3 have totals of 14 and 41. If we took the larger of these as our calculated value it would be greater than the required value (31.5) and we would reject the null hypothesis. This would lead us to the **opposite** conclusion that we reached when using the rankings in Table 6.2. Clearly it is intolerable that the result of a significance test should depend on which of two arbitrary conventions was observed when ranking the results. This ambiguity does not arise in practice because Table ST10 is intended to be used with the rankings in Table 6.2 and not with the rankings in Table 6.3. You will always have the appropriate rankings if you observe the following rule:

The calculated value of Wilcoxon's test is equal to the greater of the two rank sums. However, we must ensure that the ranks are assigned in such a way as to give maximum rank sum to the sample containing the smaller number of items.

If the two samples are equal in size the above rule can be ignored. Thus if the agricultural researcher had planted five plots with each variety, instead of 6 and 4, we would have got the same calculated value regardless of which ranking convention we used.

Exercise 6.1

In Chapter 4 we discussed an experiment which was carried out in order to assess the effectiveness of an oil additive. Petrol consumptions in miles per gallon of the 12 cars are reproduced in Table 6.4.

Table 6.4 — Petrol consumptions (mpg)

With additive						
Car	A	B	C	D	E	F
Consumption	23.7	46.8	32.1	52.7	45.2	35.0

Without additive						
Car	G	H	I	J	K	L
Consumption	48.2	21.4	36.2	43.7	26.5	30.4

In Chapter 4 we used the two-sample *t*-test to check the effect of the additive. If, however, we suspected that the variation in mpg from car to car did not follow a normal distribution, we would be wise to use a rank sum test.

(a) Rank the 12 consumption figures, giving a rank of 1 to the car with **lowest** mpg. Calculate a rank total for each of the two samples.
(b) Rank the 12 consumption figures, giving a rank of 1 to the car with the **highest** mpg. Calculate a rank total for each of the two samples.
(c) Carry out a Wilcoxon rank sum test to compare the performances of cars with and without additive.

6.3 TUKEY'S QUICK TEST

The Wilcoxon rank sum test is rather involved. I would advise you to follow the instructions very carefully when doing such tests, unless you are doing them frequently. Obviously, this advice would apply equally well to several of the significance tests we discussed in previous chapters. However, such caution is not necessary when you carry out what is known as 'Tukey's quick test'. This is an alternative to the two-sample *t*-test and the Wilcoxon rank sum test. It is so simple that you will remember it very easily. You will even be able to memorize the whole table of required values.

To illustrate the use of this simple test I shall return to the petrol consumption data in Table 6.1. In order to clarify the method of obtaining the calculated value, I have drawn the double dot plot in Fig. 6.1. Whenever I examine such a diagram my

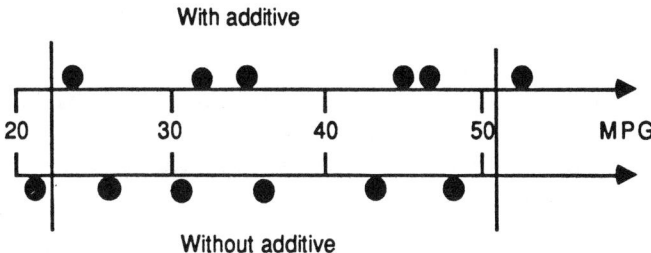

Fig. 6.1 — Tukey's quick test.

attention is drawn to the two ends, to see how many observations from one sample **protrude** beyond the other sample. In Fig. 6.1 the protrusions, or the 'excedences' as Tukey calls them, are separated from the rest of the data by the vertical lines. We can see that the total excedence is equal to 2. This is the calculated value.

Obtaining the required value is even simpler. It is equal to 7. In fact the required value is always equal to 7, regardless of the sample sizes. Why not memorize it? In fact you could easily memorize the three required values for 95%, 99% and 99.9% confidence levels; they are 7, 10 and 13 respectively.

As our calculated value, 2, is less than the required value, 7, we cannot reject the null hypothesis. The hypothesis is exactly the same as for the two-sample t-test, that 'the additive has no effect'. Thus we are unable to conclude that the additive is effective. Tukey's quick test has therefore led us to the same conclusion that we reached with the two-sample t-test and with Wilcoxon's rank sum test. You may have difficulty accepting that such a simple test could be very effective. Let me assure you, it is effective and will almost always lead us to the same conclusions as would a two-sample t-test, provided that the two samples are roughly equal in size. Tukey advises that it is unwise to use his quick test when the size of the larger sample is greater than $3+4m/3$, where m is the size of the smaller sample.

If we put m equal to 6, say, we find that $3+4m/3$ is equal to 11. Thus sample sizes of 6 and 11 would be acceptable, but 6 and 12 would not.

Tukey makes no stipulation that the two populations should have normal distributions. Nonetheless, it is clear that the conclusion given by Tukey's test could be influenced by the presence of an outlier. The following example will illustrate this point.

The data in Table 6.5 gives the tensile strength of 20 samples of synthetic rubber.

Table 6.5 — Tensile strength (kg/cm^2) of 20 rubber samples

X	236	245	276	256	254	230	245	258	239	254
Y	263	255	271	265	260	279	248	263	255	266

Ten were made with additive X and ten with additive Y. The purpose of the experiment was to ascertain which of the two additives would give the higher tensile strength.

Exercise 6.2

(a) Put the 20 tensile strength measurements into rank order. Write X or Y below each of the ranked numbers, to indicate which additive was used.
(b) From your ranked list obtain the total excedence, to use as the calculated value in a Tukey quick test.
(c) What conclusion do you draw concerning the relative effectiveness of the two additives?
(d) The sample which had a tensile strength of 276 kg/cm^2 could be regarded as a suspected outlier. Although Dixon's test does not confirm this observation to be an outlier. I suspected that it may have had considerable influence on your Tukey test in part (c). Remove the 276 and repeat Tukey's quick test, but increase the required value to 10 to compensate for the removal.

The influence of outliers on Tukey's quick test was pointed out in Neave (1979). Neave suggested a modified Tukey test in which we reject one of the observations so as to increase the total excedence as much as possible. The rejected observation will be the highest or the lowest in one of the two sets of data, as it was in Exercise 6.2. To compensate for the rejection the required values are increased from 7, 10 and 13 to 10, 13 and 16. This modification gives us a more powerful test.

6.4   THE χ^2-TEST

Dr Rathbone is a medical doctor who specializes in ear, nose and throat diseases. He has been persuaded by the Kewre Pharmaceutical Company to conduct a clinical trial in order to test their new drug Novatab. This drug has been developed in tablet form as a treatment for a particular disease which affects the inner ear. The currently recommended treatment for this disease is to administer an ointment. Altocrem, which is manufactured by a competitor of Kewre. Both treatments will be included in the trial and each of the 60 patients who have agreed to participate will be allocated at random to either Novatab or Altocrem.

The progress of this particular disease has been widely studied by medical researchers and it is generally accepted that there is a spontaneous remission rate of 43%. In other words, if all patients were left untreated, 43% of them would recover completely, while the other 57% would degenerate into an advanced stage of the disease from which an early death would be inevitable. The outcome can be predicted with near certainty within six months of the disease being diagnosed.

With this knowledge in mind Dr Rathbone decides to terminate his clinical trial after each patient has been treated for six months. At this point a decision will be made as to whether or not the patient has been cured.

During the past two years many patients being treated with Altocrem have complained of side effects which include a deepening of the voice and, more seriously, dizziness. Rathbone intends, therefore, to record any complaints of dizziness, regardless of whether the patient is on Novatab or Altocrem. The results of the clinical trial are presented in Table 6.6. (Note that this is not presented as a model clinical trial. It could be criticized for several deficiencies, including the absence of a control group.)

The only numbers in Table 6.6 are the patient numbers in the first column; these will play no part in the data analysis. In the other four columns of Table 6.6 we have four **qualitative** variables. Each of these variables gives us some information about every patient, but this information is not expressed in a numerical form. Nontheless, the contents of Table 6.6 will help us to answer the following questions:

(a) Does either treatment improve a patient's chance of recovery or would the patient be better left untreated?

(b) Is Novatab more effective than Altocrem as a treatment for patients with this disease?

(c) Is dizziness less likely to be experienced by patients on Novatab than patients on Altocrem?

(d) Is Novatab equally effective with male patients and female patients?

(e) Are male and female patients equally likely to suffer from dizziness?

All the above questions contain the word **'patients'**. It cannot be stressed too strongly that the patients referred to are not simply the 60 patients who took part in the trial but a whole population of patients. Dr Rathbone defined this **population** before the start of the trial as 'All patients in England and Wales suffering from this disease'. We shall not concern ourselves with the sampling method used by Dr Rathbone but you will be well aware, as he is, that the conclusions drawn from the data in Table 6.6 could be misleading if his sample is not representative of the population.

Let us now focus on the third of the five questions. We shall use a χ²-test (chi-squared test) to help us to decide whether or not dizziness is less likely to be experienced by patients on Novatab than by patients on Altocrem. The first step towards an answer is to count the number of times certain words occur in certain columns of Table 6.6. Counting is the only operation we can perform on qualitative data. The numbers obtained by counting are known as 'frequencies'. The relevant frequencies are gathered together in Table 6.7.

Table 6.7 is known as a **contingency table**. Alternatively, it can be described as a 'two-way frequency table' or a 'two-way contingency table'. Such tables are very useful for illustrating the relationship between two qualitative variables. In Table 6.7 the variables are 'treatment' and 'dizziness'. The two questions 'Is there a relationship between treatment asnd dizziness?' and 'Is dizziness dependent on treatment?' are equivalent to the question that we posed at the outset, 'Is dizziness less likely to be experienced by patients on Novatab than patients on Altocrem?' All three questions will be answered by the χ²-test which will proceed as follows:

(a) Firstly we shall put forward the **hypothesis** that dizziness is not dependent on treatment (i.e. a patient is equally likely to experience dizziness regardless of which treatment is given). Furthermore, we shall assume that the hypothesis is

Table 6.6 — Results of the clinical trial

Patient	Sex	Treatment	Cure	Dizziness
1	F	Novatab	Yes	No
2	F	Novatab	No	Yes
3	F	Altocrem	Yes	Yes
4	M	Novatab	Yes	Yes
5	M	Altocrem	Yes	No
6	F	Altocrem	No	Yes
7	M	Altocrem	Yes	No
8	F	Novatab	Yes	No
9	F	Altocrem	No	Yes
10	M	Novatab	Yes	No
11	M	Novatab	Yes	No
12	M	Altocrem	Yes	No
13	F	Novatab	No	No
14	M	Altocrem	Yes	Yes
15	F	Novatab	Yes	No
16	F	Altocrem	No	No
17	F	Altocrem	No	Yes
18	M	Novatab	Yes	No
19	M	Novatab	Yes	Yes
20	F	Novatab	No	No
21	F	Altocrem	No	No
22	M	Novatab	Yes	No
23	M	Altocrem	Yes	No
24	F	Altocrem	No	No
25	F	Novatab	Yes	No
26	M	Altocrem	No	No
27	F	Altocrem	No	Yes
28	M	Altocrem	No	No
29	F	Novatab	Yes	No
30	M	Altocrem	No	No
31	M	Novatab	Yes	No
32	M	Altocrem	Yes	No
33	F	Altocrem	Yes	No
34	F	Novatab	Yes	Yes
35	M	Altocrem	No	Yes
36	M	Novatab	Yes	No
37	F	Altocrem	No	Yes
38	M	Novatab	Yes	No
39	F	Novatab	Yes	No
40	F	Altocrem	No	Yes
41	F	Novatab	Yes	No
42	M	Altocrem	Yes	Yes
43	M	Altocrem	No	No
44	F	Novatab	Yes	No
45	M	Altocrem	Yes	No
46	F	Novatab	No	Yes
47	F	Novatab	No	No
48	M	Novatab	Yes	No
49	F	Altocrem	Yes	No
50	M	Altocrem	Yes	Yes
51	F	Novatab	No	Yes
52	F	Altocrem	No	No
53	M	Novatab	Yes	No
54	F	Altocrem	No	Yes
55	F	Novatab	Yes	No
56	F	Novatab	No	No
57	M	Altocrem	Yes	No
58	M	Novatab	Yes	No
59	M	Novatab	Yes	No
60	F	Altocrem	Yes	Yes

Table 6.7 — Observed frequencies

	Dizziness Yes	No	Total
Treatment			
Novatab	6	24	30
Altocrem	12	18	30
Total	18	42	60

true until it is proved false.

(b) Then we shall calculate the frequencies that we would expect to obtain in Table 6.7 if the hypothesis were true. These are known as **expected frequencies**.

(c) We shall compare these expected frequencies with the observed frequencies in Table 6.7.

(d) If the expected frequencies differ considerably from the observed frequencies we shall decide that the assumption made in (a) is contradicted by the data and we shall conclude that dizziness is dependent on treatment.

If we start with the assumption that 'the experience of dizziness does **not** depend on which treatment is prescribed', what frequncies would we expect to find in Table 6.7? Surely it is reasonable to suggest that the 18 patients in the bottom row would be split equally into two groups of 9 and the 42 would be split equally into two groups of 21. This would give the expected frequencies in Table 6.8.

Table 6.8 — Expected frequencies

	Dizziness Yes	No	Total
Treatment			
Novatab	9	21	30
Altocrem	9	21	30
Total	18	42	60

Note the similarity between Tables 6.7 and 6.8. They have the same column totals (18 and 42), the same row totals (30 and 30) and the same overall total (60). The calculation of expected frequencies was very easy with these data because Dr Rathbone had allocated equal numbers of patients to each drug. If 40 patients had received Novatab while only 20 had been treated with Altocrem the expected frequencies would not be so obvious. Fortunately the following formula can be used in all cases.

Expected frequency=(column total) (row total)/(overall total)

Thus for the top left corner of the table (i.e. the patients on Novatab who did not experience dizziness) we calculate:

$$\text{expected frequency} = (18) \ (30)/(60)$$
$$= 9$$

Before we compare the observed and expected frequencies, which is the next step in the procedure, I would like to ensure that you know exactly how they are obtained.

Exercise 6.3

I would like you to answer Dr Rathbone's fourth question: 'Is Novatab equally effective with male patients and female patients?' Thus in Table 6.6 you will need to concentrate on the columns headed 'Sex' and 'Cure', but you will also need to refer to the 'Treatment' column in order to identify the patients on Novatab. Many people experience some difficulty inspecting three columns simultaneously.

(a) Draw a contingency table (like Table 6.7 in the text) into which you can put observed frequencies.
(b) Using Table 6.6 count the observed frequencies and complete your contingency table.
(c) Draw a second table exactly like the one in part (a) and insert the same row totals and column totals but leave the centre blank to accommodate expected frequencies.
(d) Use the formula given in the text to calculate expected frequencies and complete the second table (round your expected frequencies to one decimal place).
(e) Compare your two tables and answer the question posed at the start of this exercise.

We are now ready to make a formal comparison of the observed and the expected frequencies. One way to do this would be to subtract each expected frequency in Table 6.8 from the corresponding observed frequency in Table 6.7 then add up the differences. However, common sense would suggest that we are likely to get bigger differences if we have larger frequencies and it seems desirable that the size of the frequencies should be taken into account. Mathematical statisticians have considered various possibilities and concluded that the best method of comparison is to obtain a calculated value using the formula:

For a χ^2-**test** the calculated value is obtained using the formula:

$$\text{calculated value} = \Sigma[(O-E)^2/E]$$

where O=observed frequency and E=expected frequency.

We carry out the calculation of $(O-E)^2/E$ for each of the four cells in the contingency table and add the results. Using the frequencies in Tables 6.7 and 6.8 we obtain

$$\text{Calculated value} = [(6-9)^2/9] + [24-21)^2/21] + [(12-9)^2/9] + [(18-21)^2/21]$$
$$= 1.0 + 0.429 + 1.0 + 0.429$$
$$= 2.858$$

To obtain a required value from Table ST11 we need to know how many degrees of freedom to use. Surprisingly this is not dependent on the number of people in the sample but is dictated by the shape of contingency table. A table with 2 rows and 2 columns, as we have in Tables 6.7 and 6.8, will always have 1 degree of freedom, regardless of the sample size. The degrees of freedom for this type of test can be calculated from the following formula:

For a χ^2-test on a contingency table the degrees of freedom are given by:
degrees of freedom=(number of rows−1) (Number of columns−1)

Thus for a contingency table with two rows and two columns we have degrees of freedom=(2−1) (2−1)=1.

Using 1 degree of freedom and 95% confidence we find a table value of 3.84 from table ST11. Thus our calculated value (2.858) is less than the required value (3.84) so we cannot reject the hypothesis. We cannot reasonably conclude, therefore, that the experience of dizziness is dependent on which treatment is given. In other words we cannot conclude that dizziness is more prevalent with one drug than with the other.

Exercise 6.4

Let us return to the question posed in Exercise 6.3: 'Is Novatab equally effective with male and female patients?' In Exercise 6.3 you drew up a table of observed frequencies and calculated expected frequencies. You are now invited to make a formal comparison and to draw a conclusion.

(a) Use the observed and expected frequencies from Exercise 6.3 to obtain a calculated value.
(b) Obtain a required value from Table ST11.
(c) Compare the calculated value with the required value and make a decision regarding the truth of the hypothesis 'Novatab is equally effective with male and female patients'.

6.5   CONFIDENCE LIMITS FOR POPULATION PERCENTAGES

As I asserted earlier in this chapter, the only operation we can perform on qualitative data is to **count**. In fact qualitative data are sometimes referred to as 'counted data'. Having counted how many items in our sample fall into a particular category it may be very useful to convert the count into a percentage. For example, we see in Table

6.6 that 23 of the 30 patients on Novatab were cured. Expressing 23 as a percentage of 30 we obtain 76.7%. This would appear to indicate the Novatab is an effective treatment, for 76.7% is much greater than the spontaneous remission rate of 43%.

Before you rush out to the pharmacy, let me remind you that the 76.7% is based on a **sample** of patients, which might not be representative of the population. Furthermore, let me remind you that Table 6.6 contains qualitative data. We shall see later that samples need to be larger when we gather qualitative data than when we obtain the quantitative data with which you are probably more familiar. Let us calculate confidence limits for the percentage of patients that would have been cured if we had treated with Novatab the whole population of sufferers. The width of this interval will convince you that sample sizes need to be large when we are gathering qualitative data.

Confidence limits for a population percentage are given by

$$p \pm \{k\sqrt{[p(100-p)/n]+(50/n)}\}$$

where p is the sample percentage, n is the sample size and k is equal to 1.96 for 95% confidence and 2.58 for 99% confidence.

Note: the above formula should only be used with a sample size of 30 or more.

The values of k given above are taken from the normal distribution. The $50/n$ in the formula is known as a 'continuity correction'. It is included because strictly speaking the normal distribution is applicable only to continuous data, while we are using it with counted data. Substituting $k=1.96$, $p=76.7$ and $n=30$ we obtain the following confidence limits:

$$76.7 \pm \{1.96\sqrt{[76.7(100-76.7)/30]+(50/30)}\}$$
$$=76.7 \pm 16.8$$
$$=59.9\% \text{ to } 93.5\%$$

Thus we can be 95% confident that, if all patients were treated with Novatab, the percentage cured would lie between 59.9% and 93.5%. The confidence interval is rather wide. However, it does not include 43%, the spontaneous remission rate, so we can conclude that Novatab is effective.

Exercise 6.5

(a) Using the data in Table 6.6 calculate the percentage of the 30 patients on Altocrem who were cured.

(b) Calculate 95% confidence limits for the percentage cure if all sufferers were treated with Altocrem.

(c) Compare the confidence limits calculated in part (b) with those for the cure percentage of Novatab. Can we reasonably conclude that Novatab is more effective than Altocrem?

The fact that the two confidence intervals overlap suggests that the two population percentages **could** be equal, which implies that the two drugs **could** be equally effective. However, this line of reasoning is likely to lead us to a false conclusion, as we saw in Chapter 2 when discussing confidence intervals for population means. Let me repeat the argument that I put forward on that occasion. The Altocrem percentage could be as high as 66.2% but it is probably well below this upper limit. Similarly the Novatab percentage could be as low as 59.9%, but it is probably well above this lower limit. It is therefore very unlikely that the two percentages will be equal, even though the intervals overlap to some extent.

In Chapter 2 we overcame this difficulty by calculating confidence limits for the difference between the two population means. We shall now calculate confidence limits for the difference between the two population percentages.

Confidence limits for the difference between two population percentage are given by:

$$(p_1-p_2)\pm k\sqrt{\{p(100-p)[(1/n_1)+(1/n_2)]\}}$$

where p_1 is the larger of the two sample percentages, p_2 is the smaller of the two sample percentages, p is the combined sample percentage, n_1 is the size of the sample which gave us p_1, n_2 is the size of the sample which gave us p_2 and k is equal to 1.96 for 95% confidence and 2.58 for 99% confidence.

From the data in Table 6.6 we have $p_1=76.7$, $p_2=46.7$, $n_1=30$ and $n_2=30$. As 37 of the 60 patients were cured the combined sample percentage is $(37/60)100$ which is 61.7%. For 95% confidence the value of k from the normal distribution is 1.96. Confidence limits for the difference between the two population percentages are:

$$(p_1-p_2)\pm k\sqrt{\{p(100-p)[(1/n_1)+(1/n_2)]\}}$$
$$=(76.7-46.7)\pm 1.96\sqrt{\{61.7(38.3)[(1/30)+(1/30)]\}}$$
$$=30.0\pm 24.6$$
$$=5.4\% \text{ to } 54.6\%$$

Thus we can be 95% confident that the difference between the two population percentages lies between 5.4% and 54.6%. What does this mean in practical terms? As the interval **does not include zero** we can reasonably conclude that the two drugs differ in their effectiveness. Since Novatab gave the higher cure rate (76.7% compared with 46.7%) we conclude that Novatab is superior to Altocrem. Furthermore, we can say that the use of Novatab, rather than the currently used Altocrem, will result in the percentage of patients being cured increasing by at least 5.4% and possibly as much as 54.6%.

We have now answered three of the five questions posed by Dr Rathbone. We have just concluded that Novatab is more effective than Altocrem. Earlier we concluded that the use of Novatab improved a patient's chance of recovery. We shall now reconsider the question we answered in Section 6.4 using a χ^2-test: 'Is dizziness less likely to be experienced by patients on Novatab than patients on Altocrem?' This question could also be answered by calculating confidence limits for the difference between two population percentages.

Exercise 6.6

(a) Turn to Table 6.6 and count:
   (i) the number of patients on Novatab who experienced dizziness;
   (ii) The number of patients on Altocrem who experienced dizziness.
(b) For each treatment calculate the percentage of patients who experienced dizziness. Calculate the combined sample percentage.
(c) Use the three percentages from (b) to calculate 95% confidence limits for the difference between the population percentages.
(d) What conclusions can you draw from the confidence limits calculated in part (c)?

6.6 LARGER SAMPLES ARE NEEDED WITH QUALITATIVE DATA

You may have been surprised by the width of the confidence intervals calculated in this chapter. For example, we calculated confidence limits for the cure rate of the drug Novatab. The interval was $76.7\% \pm 16.8\%$. It may have surprised you to obtain such a wide interval with a sample size of 30.

Compare this confidence interval with what you might have found if you were estimating the mean height of patients, using a sample of 30. Suppose, for example, the sample mean height were 65 inches and the sample standard deviation were 5 inches. Using $\bar{x}=65$, $s=5$ and $n=30$ we obtain a confidence interval of 65 ± 1.9 inches which is relatively narrower than $76.7\% \pm 16.8\%$. It is unfortunately true that **qualitative** data are likely to give a much wider confidence interval than we would find with **quantitative** data from a sample of the same size.

There are, however, many situations in which researchers gather qualitative data to estimate percentages and yet they require narrow confidence intervals. Consider, for example, an opinion poll to determine voters' intentions in a forthcoming election. It would be little use reporting that the percentage of electors who intended to vote Conservative was $30\% \pm 16\%$. However, this is the sort of result we might obtain with a sample of 30 electors. Obviously a much larger sample is required, but how large would the sample need to be give $\pm1\%$, say?

It is much easier to answer such questions if we simplify our confidence interval formula. You will recall that we calculated confidence limits for a population percentage using

$$p \pm k\sqrt{\{[p(100-p)/n]+(50/n)\}}$$

If we abandon the continuity correction $(50/n)$ and let $k=2$ and let $p=50$ inside the square root, then we obtain the following:

Approximate 95% confidence limits for a population percentage are given by:

$$p\pm100/\sqrt{n}$$

Note: this formula should not be used if your sample size is less than 100.

You might be a little alarmed by the cavalier manner in which the formula was simplified. However, if you use the simplified formula you may be surprised by its close agreement with the original. Our immediate use for the simplified formula is to demonstrate how large a sample is needed in order to provide a sufficiently narrow confidence interval. To achieve this we simply substitute various values of n into the formula to obtain the widths in Table 6.9.

Table 6.9 — Width of confidence interval for population percentage

Sample size	Approximate confidence interval
100	$p\pm10\%$
400	$p\pm5\%$
1000	$p\pm3\%$
2500	$p\pm2\%$
10000	$p\pm1\%$

We can see in Table 6.9 that you need a sample size of 10 000 if you wish to estimate a population percentage to within 1%. Very rarely do opinion pollsters use such a large sample and voting intentions are often assessed using samples between 500 and 2000.

The random sampling assumption is equally relevant whether we are dealing with qualitative or with quantitative data. Thus the pollster needs a random sample of 10 000 people in order to estimate a population percentage to within $\pm1\%$. If a biased method of sampling is used the effect of increasing the sample size is to focus more precisely on a percentage that is false.

6.7 SIGNIFICANCE LEVELS

Throughout Chapters 1–6 we have used 95% confidence whenever we have referred to a statistical table. You may have wondered why these tables contain other levels of confidence if no one ever refers to them. The simple truth is that the scientist does **not** automatically use 95% confidence on every occasion, but selects a percentage which is appropriate for the situation. Furthermore, many scientists do **not** use the expressions 'level of confidence' or 'confidence level'. They prefer to speak of **significance level**:

Significance level = 100% − confidence level

Obviously we have been using a 5% significance level throughout Chapters 4, 5 and 6. If we had referred to the 99% column, rather than the 95% column of the statistical tables, we would have been using a 1% significance level. Had we used the even higher confidence level of 99.9% the significance level would have been only 0.1%.

Many scientists prefer to speak of 'significance level' rather than 'confidence level' because it is more directly related to the risks that we run whenever we carry out a significance test. **If we use a 5% significance level there is less than 5% chance of rejecting a null hypothesis which is true**. If we use a 1% significance level this risk is reduced to less than 1%, or 1 in 100. We noted in Chapter 4 that a small calculated value is consistent with the null hypothesis, while a large calculated value constitutes evidence against the null hypothesis. This point can be expressed quantitatively as follows. If the null hypothesis is true there is only a 5% chance (i.e. 1 in 20) of obtaining a calculated value greater than the required value in the 95% confidence column. Furthermore, there is only 1% chance of obtaining a calculated value greater than the required value from the 99% confidence column. If an experiment were repeated many, many times and a calculated value obtained on each occasion, the distribution of the calculated values might resemble Fig. 6.2.

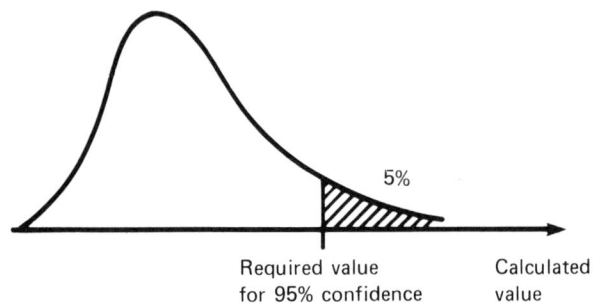

Fig. 6.2 — Calculated values from an experiment repeated many times (in a situation where the null hypothesis is true).

Regardless of whether you speak of 'confidence level' or 'significance level' the conclusion you draw from a test may well depend on the percentage you decide to use. Thus you will want to know how a scientist chooses a suitable percentage and what are the implications of his choice. Perhaps these rather difficult ideas will be clarified if we examine a specific example.

In Chapter 5 Dr Chewsey was involved in preference testing of two shampoos. He asked 50 people to try both products and 35 expressed a preference for brand X. Using a calculated value of 35 and a required value of 32.5 you drew the conclusion that brand X was preferred by a majority of the population. What conclusion would you have reached if you had used some other level of confidence? To help us to answer this question the table values from Table ST8 are reproduced in Table 6.10.

Table 6.10 — Values from table ST8 for a sample size of 50

Confidence level	90%	95%	98%	99%	99.8%	99.9%
Significance level	10%	5%	2%	1%	0.2%	0.1%
Required value	31.5	32.5	33.5	34.5	36.5	36.5

With a calculated value of 35, the conclusion we draw will depend on whether the required value is greater than or less than 35. Thus we would reject the null hypothesis if we were using a significance level of 10%, 5% or 1%. However, we would not reject the null hypothesis if we were using 0.2% or 0.1%. Obviously Dr Chewsey can modify his conclusion by changing his significance level. His scope for doing so is illustrated in Table 6.11.

Table 6.11 — The three possible conclusions

Conclusion	Significance level
X is preferred to Y	This is the appropriate conclusion when using 10%, 5%, 2% or 1% significance
We are unable to conclude that X and Y differ	This is the appropriate conclusion when using 0.2% or 0.1% significance
Y is preferred to X	This conclusion cannot reasonably be drawn no matter what significance level is used

We can see in Table 6.11 that Dr Chewsey does not have complete freedom of choice among the three possible conclusions. He could not reasonably conclude that brand Y is preferred to brand X, because only 15 of the 50 assessors chose brand Y. However, he can adopt either of the other two conclusions depending on the level of significance that he uses.

Many scientists are horrified when they discover that the significance level can be manipulated in order to change the conclusions drawn from a set of data. They do not wish to be associated with any practice which is so arbitrary, so subjective or so malleable. Perhaps it is because of significance level manipulation that non-statisticians speak of '... lies, damn lies, and statistics'.

Many scientists realize, however, that significance tests and confidence intervals are essential tools of their trade. How then can the scientists make use of these techniques without compromising his/her reputation as an honest, decent and dependable member of society? One answer is to choose the level of significance before data analysis commences. Perhaps a better answer is to leave the choice of significance level to someone else. We shall explore this possibility when we discuss the reporting of significance tests, later in this chapter. First we must examine another contentious aspect of significance testing.

6.8   ONE-SIDED AND TWO-SIDED TESTS

In the previous section it was revealed that the outcome of a significance test may depend on the choice of significance level. It is unlikely that you received this news with enthusiasm. However, there is more bad news to follow. It must now be revealed that the conclusion reached from a significance test may also depend on whether you use a two-sided test or a one-sided test. All the tests that we have carried out so far have been two-sided tests. They have all shared the feature which stamps a test as two-sided — they have all had **three possible conclusions**.

In Table 6.11 are listed the three conclusions that could be drawn from the yes–no test. Thus we might conclude that 'X is preferred to Y' or we might conclude that 'Y is preferred to X' or we might conclude that 'there is no significant difference between the two products'. The conclusion that we draw will be dictated by the results of the experiment, of course. However, until the experiment has been carried out any one of the three conclusions is possible.

Suppose that Dr Chewsey is interestred in only two of the three possibilities. This would be the case if X were a new product and Y was the established brand. The new product (X) would only be adopted if it proved itself superior to the established product (Y). Thus the purpose of the experiment would be to prove that X was superior to Y; and the significance test would be carried out **only if a majority of the 50 participants favoured brand X**. From the outset, therefore, we know that only two conclusions are possible and we therefore carry out a one-sided test:

A one-sided yes–no test

Null hypothesis	50% of the population prefer brand X and 50% prefer brand Y.
Calculated value	=The number of people who prefer brand X =35
Required value	=31.5 from the **90%** column of Table ST8 for a sample size of 50.
Conclusion	Because the calculated value is greater than the required value we reject the null hypothesis and conclude that more than 50% of the population prefer brand X.

Note the differences between this one-sided test and the two-sided test carried out in Chapter 5. The tests differ in two respects:

(a) the instructions for obtaining the calculated value;
(b) the column of Table ST8 from which we obtain the required value.

Because, in a one-sided test, we entertain only two of the three possibilities the risk of a wrong conclusion is reduced. In fact it is halved. To achieve a 5% significance level with a one-sided test we must, therefore, use the **90%** column. This adjustment is illustrated in Table 6.12.

Table 6.12 — The use of tables for one- and two-sided tests

Confidence level given in the table	90%	95%	98%	99%	99.8%	99.9%
Significance level for a two-sided test	10%	5%	2%	1%	0.2%	0.1%
Significance level for a one-sided test	5%	2.5%	1%	0.5%	0.1	0.05%

You may find this discussion of one-sided and two-sided tests very difficult to follow. This is because we have not discussed significance testing in sufficient depth for you to appreciate the finer points of the argument. Further discussion, however, would elevate the topic to a status it does not deserve. If you wish to acquire a thorough grasp of the distinction between one-sided and two-sided tests you must refer to other texts. Before doing so, however, ask yourself whether it is worth the time and effort that will be needed to gain this extra understanding. Why not adopt the rule 'if in doubt do a two-sided test'? The worst that can happen is that the actual risks involved are not quite what you thought they were.

An even more important rule to observe is 'whatever test you carry out always record in your report exactly what you did'. If you follow this rule you cannot be accused of dishonesty.

6.9 REPORTING SIGNIFICANCE TESTS

In the last two sections we have discussed significance levels and one-sided tests. You must now be aware that significance testing is not the innocent and carefree activity that it first appeared. You may even be haunted by visions of a scientist who carries out perfect experiments then suffers moral torment during data analysis.

Let me help you to get significance testing into perspective. You may be more willing to make use of significance tests after you have considered the following points:

(a) Important decisions are rarely based on one set of data in isolation.

(b) The person who makes an important decision is probably not the person who carried out the experiment or the person who analysed the data.

Obviously the scientist needs a method of **reporting** the results of his experiments to a wider audience. Readers of this report, who may include people of higher authority, can integrate his findings into other available knowledge to reach conclusions or to make decisions. When we look at significance testing in this context it is clear that the reader of the report is the best person to choose the significance level. Furthermore the reader may be in a better position to say whether a one-sided test or a two-sided test is appropriate.

With these thoughts in mind let us return to some of the significance tests carried out in this and earlier chapters and see how they might be reported. We shall start with the yes–no test in Chapter 5. The result of this test could be reported as follows:

35 of the 50 assessors expressed a preference for brand X. A yes–no test indicates that a majority of potential customers would prefer this brand ($p < 0.01$, two-sided).

The writer of the report has told us which type of test was carried out and that the finding is significant at the 1% level with a two-sided test. The expression '$p < 0.01$' is read as 'p less than nought point nought one'. The p is short for probability and $p < 0.01$ indicates that there is less than 1% chance of the conclusions being wrong.

We shall now turn to a paired comparison test that was used to assess the effectiveness of a diet. The data were introduced in Exercise 4.7 and referred to again in Chapter 5. The result of this paired comparison test could be reported as follows:

The mean weight loss of the eight people on the diet was 4.25 lb. A paired comparison test indicates that the diet is effective ($t = 3.04$, df$=7$, $p < 0.02$, two sided).

The researcher is informing us in his report that a paired comparison test was carried out. Perhaps he assumes that readers of his report will be familiar with this test. If he is in doubt he could give a reference to a text in which the paired comparison test is described, e.g. Caulcutt (1983). The researcher is also telling us that the calculated value is equal to 3.04 and that 7 degrees of freedom were used to obtain the required values. Finally we are informed that the mean weight loss is significant at the 2% level if a two-sided test is used.

A reader of the report may consider that a one-sided test would have been more appropriate. It is reasonable to argue that the researcher would **not** have carried out a test if the people on the diet had shown an increase in weight on average. Thus it is fair to halve the quoted percentage and claim that the weight loss is significant at the 1% level. Perhaps the report would be better if it were amended to read '($t = 3.04$, df$= 7$, $p < 0.01$, one sided)'.

For a final example let us reconsider the χ^2-test carried out in section 6.5. The purpose of the test was to answer the question 'Is Novatab equally effective with male and female patients?' The report might read as follows:

All the 13 male patients were cured but only 10 of the 17 female patients were cured. A χ^2-test shows that Novatab is more effective with male patients than with females ($\chi^2=6.842$, df$=1$, $p<0.01$, two sided).

The scientist tells us in his report that a χ^2-test was carried out and the calculated value was equal to 6.842. Required values with 1 degree of freedom were compared with this test value and it was found that the difference between males and females was significant at the 1% level. It is entirely appropriate to use a two-sided test as the scientist would have proceeded with the test regardless of whether males or females were found to have the higher cure rate.

6.10   A SUMMARY OF THE IMPORTANT POINTS

(1) In this chapter I have introduced three significance tests to add to the eight covered in Chapters 4 and 5.
(2) **Wilcoxon's rank sum test** is a non-parametric alternative to the two-sample t-test. It is not restricted by the assumption that the population has a normal distribution.
(3) **Tukey's quick test** is also an alternative to the two-sample t-test. With this remarkably simple test the calculated value is the total excedence and the required values are 7, 10 and 13.
(4) The χ^2**-test** is used to check the association between two qualitative variables. (Note that there is a 'χ^2 goodness-of-fit test', which is not covered in this text.)
(5) With **qualitative data** we cannot perform the arithmetic operations that we use with quantitative data. However, we can count how many items fall into each category and we can express the counts as percentages. Sample percentages can be used to calculate confidence limits for population percentages.
(6) Confidence limits for the difference between two **population percentages** can help us to compare two treatments when the response is a qualitative variable. If the confidence interval does **not** include zero we can conclude that the two treatments differ. This approach is an alternative to the χ^2-test.
(7) Many scientists prefer to speak of 5% **significance level**, rather tha 95% confidence level. If you use the 5% significance level when doing a significance test you run a 1 in 20 risk of rejecting the null hypothesis, if it is true.
(8) The distinction between **one-sided** and **two-sided** significance tests is a great worry to some users of statistics. If in doubt do a two-sided test. However, there are situations in which it is right and proper to do a one-sided test.
(9) The conclusion you reach when you do a significance test depends on your choice of significance level and your choice of one-sided or two-sided test. If you **report** the outcome of your significance test in the manner I have suggested you can leave these subjective choices to the reader of your report.

6.11   ADDITIONAL EXERCISES

Exercise 6.7

George Banks is the Quality Manager of Parker Ceramics, a company which manufactures electrical insulators. With ceramic insulators intended for outdoor use the surface finish of the insulators is extremely important. During manufacture the surface coating is formed by passing the insulators through a continuous belt furnace. Every insulator is visually inspected as it leaves the furnace to detect 'hazing' of the surface. Past experience shows that, even when the furnace is working at its best, we can expect that 12% of insulators will be hazed.

During one particular shift 32 insulators are produced in the first hour and 8 of these are found to be hazed. Even though the defective rate is 25%, George Banks decides that no adjustment should be made at this time. However, during the second hour a further 48 insulators are produced and these contain 10 which are hazed. Although the defective rate has decreased George gives instructions for production to stop whilst an adjustment is made. During the remaining six hours of the shift only 25 hazed insulators are found in the 220 produced.

(a) Regard the 32 insulators produced in the first hour as a random sample of what might be produced later if no adjustment is made. Use the data from the first hour to calculated 95% confidence limits for the percentage of insulators that will be hazed if no adjustment is made. Did George Banks make the correct decision at the end of the first hour?

(b) Repeat part (a) using all the data for the first two hours. Did George Banks make the correct decision at the end of the second hour?

(c) (Harder) If George Banks were using the procedure you have followed in parts (a) and (b), what would be the risks, at each decision, of his stopping the process even though it was producing only 12% hazed insulators?

(d) Compare the data for the first two hours with those for the last six hours, to decide whether or not the adjustment was successful in reducing the percentage of hazing. (In this chapter I have introduced two methods you could use. Why not try both?)

(e) In this situation is it appropriate to use one-sided or two-sided significance tests?

Exercise 6.8

John Davenport has been asked to investigate the causes of recent fractures in certain joints of the new pop-up toaster. He suspects that the bond strength might depend on the way that the adhesive is applied and he carries out a bench experiment to compare two methods of application, centre spot and diagonal stroke. The strengths of eight joints made by each method are given in Table 6.13.

Table 6.13 — Bond strength (kg) with two methods of application

Centre spot				Diagonal stroke			
26.5	23.1	25.7	26.7	25.2	31.2	28.0	27.0
23.9	21.5	30.0	24.2	28.5	29.0	33.1	27.6

(a) Name three types of significance test that could be used to decide whether or not the diagonal stroke methods gives higher bond strength than the centre spot method.
(b) Carry out each of the three significance tests and advise John Davenport on the bond strengths given by the two methods.
(c) Comment on the relative merits of the three tests you carried out in part (a).
(d) Would it be reasonable of me to suggest that you should feel shame, disgust and guilt if you have proceeded to this point, without drawing a dot plot of the data?

6.12 WORKED SOLUTIONS

Solution to Exercise 6.1

(a)

Table 6.14 — Solution to Exercise 6.1(a)

Car	A	B	C	D	E	F		G	H	I	J	K	L
mpg	23.7	46.8	32.1	52.7	45.2	35.0		48.2	21.4	36.2	42.7	26.5	30.4
Rank	2	10	5	12	9	6		11	1	7	8	3	4
Total			44							34			

(b)

Table 6.15 — Solution to Exercise 6.1(b)

Rank	11	3	8	1	4	7		2	12	6	5	10	9
Total			34							44			

We can see from the rank totals in parts (a) and (b) that the larger of the two rank totals is equal to 44 regardless of which convention we use. Remember that this is only true when the two samples are equal in size. If we have unequal sample sizes we must use the convention which gives the maximum rank sum to the smaller sample.

(c)

Null hypothesis	The two population medians are equal (i.e. the additive has no effect on petrol consumption).
Calculated value	=The larger of the two rank sums =44
Required value	=51.5 ... from Table ST10 with sample sizes of 6 and 95% confidence.
Conclusion	Because the calculated value is less than the required vaslue we cannot reject the null hypothesis. Thus we are unable to conclude that the additive has any effect on petrol consumption.

This conclusion is in agreement with that reached when we carried out at a two-sample t-test in Chapter 4.

Solution to Exercise 6.2
(a)   230 236 239 245 245 248 254 254 255 255
        X   X   X   X   X   Y   X   X   Y   Y

      256 258 260 263 263 265 266 271 276 279
        X   X   Y   Y   Y   Y   Y   Y   X   Y

(b) Total excedence=5+1=6.
(c) As the calculated value, 6, is less than the required value, 7, we cannot reject the null hypothesis. Thus we are unable to conclude that the two additives differ in their effect on the tensile strength.
(d) After removing the 276 from the list of ranked data we see that the total excedence is equal to 5 plus 7, which is 12. As this calculated value exceeds the required value of 10, we can reject the null hypothesis and conclude that the two additives do differ.

Solution to Exercise 6.3
(a), (b)

Table 6.16 — Observed frequencies for Exercise 6.3.

	Cure		
	Yes	No	Total
Sex			
Male	13	0	13
Female	10	7	17
Total	23	7	30

(c), (d)

Table 6.17 — Expected frequencies for Exercise 6.3

	Cure		
	Yes	No	Total
Sex			
Male	10.0	3.0	13
Female	13.0	4.0	17
Total	23	7	30

(e) When we examine the table of observed frequencies we see that **all** the males were cured whereas only 10 of the 17 females were cured. We are tempted to conclude that Novatab is more effective with males than females. However, when we compare the two tables we realize that each observed frequency differs from the corresponding expected frequency by only 3.0. Thus if three of the males had not been cured there would be no difference at all. Now we are not sure that Novatab is more effective with males. Perhaps we need a larger sample.

Solution to Exercise 6.4

(a) Calculated value$=\Sigma[(O-E)^2/E]$
$$=[(13-10.0)^2/10.0]+[(0-3.0)^2/3.0]]$$
$$+[(10-13.0)^2/13.0]+[(7-4.0)^2/4.0]$$
$$=0.900+3.000+0.692+2.250$$
$$=6.842$$

(b) Degrees of freedom$=$(number of rows -1) (number of columns -1)
$$=(2-1)$$
$$=1$$

Using 1 degree of freedom and 95% confidence we find a required value from Table ST11 of 3.84.

(c) As the calculated value is greater than the required value we can reject the hypothesis and conclude that Novatab is **not** equally effective with males and females. We can see from the table of observed frequencies in Exercise 6.3 that Novatab is more effective with males than with females

Solution to Exercise 6.5

(a) 14 of the 30 patients on Altocrem were cured. Expressing 14 as a percentage of 30 we obtain 46.7%.

(b) Using $p=46.7$, $n=30$ and $k=1.96$ we find the following confidence limits:

$$p\pm\{k\sqrt{[p(100-p)/n]+(50/n)}\}$$
$$=46.7\pm\{1.96\sqrt{[46.7(100-46.7)/30]+[(50/30)]}\}$$
$$=46.7\pm19.5$$
$$=27.2\% \text{ to } 66.2\%$$

Thus we can be 95% confident that, if all patients were treated with Altrocrem, the percentage cured would lie between 27.2% and 66.2%. As this confidence interval includes 43% we have failed to prove that Altocrem is effective.

(c) You will have noticed that the two confidence intervals overlap. Thus it is possible that the two treatments could be equally effective. You may recall, however, that we discussed the overlap of confidence intervals in Chapter 2, and concluded that a better approach was available. I shall repeat the argument when you return to the main text.

Solution to Exercise 6.6

(a) (i) 6 of the patients on Novatab experienced dizziness.
     (ii) 12 of the 30 patients on Altocrem experienced dizziness.
(b)   (6/30)100 $=20\%$
               (12/30)100$=40\%$

Thus 20% of the patients on Novatab experienced dizziness while 40% of the

patients on Altocrem experienced dizziness.
(c) $p_1 = 40\%$, $p_2 = 20\%$, $n_1 = 30$, $n_2 = 30$ and $p = 30\%$. For 95% confidence $k = 1.96$: therefore

$$(p_1 - p_2) \pm k\sqrt{\{p(100 - p)[(1/n_1) + (1/n_2)]\}}$$
$$= (40 - 20) \pm 1.96\sqrt{\{30(70)[1/30) + /1/30)]\}}$$
$$= 20.0 \pm 23.2$$
$$= -3.2\% \text{ to } 43.2\%$$

(d) As the confidence interval in part (c) contains zero we are unable to conclude that dizziness is more prevalent with one drug than with the other. There is certainly some evidence that the percentage of Novatab patients experiencing dizziness (20% of the sample) is less than the percentage of Altocrem patients reporting dizziness (40% of the sample). However, this evidence is inconclusive.

Solution to Exercise 6.7
(a) Sample percentage $= (8/32)100 = 25\%$. The 95% confidence limits for the population percentage are:

$$p \pm 1.96\sqrt{\{[p(100 - p)/n] + (50/n)\}}$$
$$= 25 \pm 1.96\sqrt{\{[25(75)/32] + (50/32)\}}$$
$$= 25 \pm 15.2$$
$$= 9.8\% \text{ to } 40.2\%$$

Thus we can be 95% confident that the percentage of hazed insulators will lie between 9.8% and 40.2% in the long run if no adjustment is made. (This statement implies that there is a stable population percentage. If the furnace were prone to gradual continuous change, or drift, then this stability would not exist and this method of analysis would not be appropriate.) As the confidence interval includes 12% we have failed to prove that an adjustment is necessary. This would support George Bank's decision not to adjust the process.
(b) Sample percentage $= [(8 + 10)/(32 + 48)] \times 100 = 22.5\%$. The 95% confidence limits for the population percentage are given by:

$$p \pm 1.96\sqrt{\{[p(100 - p)/n] + (50/n)\}}$$
$$= 22.5 \pm 1.96\sqrt{\{[22.5(77.5)/80] + (50/80)\}}$$
$$= 22.5 \pm 9.2$$
$$= 13.2\% \text{ to } 31.8\%$$

As this confidence interval does not include 12% we have proved beyond doubt that the furnace is not performing as well as it might. This would support George Banks' decision to adjust the process at the end of the second hour.
(c) The procedure you have followed in parts (a) and (b) is equivalent to carrying out a significant test using a 5% significance level and with the null hypothesis 'the hazing rate is equal to 12%'. At each decision George Banks runs a 5% risk of rejecting this hypothesis, if it is true. Thus he runs a 1 in 20 risk of stopping the process when no adjustment is needed. He also runs a second risk, the risk of failing to stop the process even though an adjustment is needed. It is more difficult to quantify this second risk and it has not been mentioned in this

chapter. However, George Banks must balance these two risks, using very little data, and without the benefit of any training in statistics.

(d) You could use a χ^2-test, or you could calculate confidence limits for the difference between two population percentages. I shall do both.

Table 6.18 — Observed and expected frequencies

| | Hazed | | |
	Yes	No	Total
Time			
Before	18 (11.5)	62 (68.5)	80
After	25 (31.5)	195 (188.5)	220
Total	43	257	300

Null hypothesis	The population percentages before and after adjustment are equal (i.e. the adjustment had no effect).
Calculated value	$= \Sigma[(O-E)^2/E]$ $= [(18-11.5)^2/11.5] + [(62-68.5)^2/68.5$ $\quad + [(25-31.5)^2/31.5] + [(195-188.5)^2/188.5]$ $= 3.67 + 0.62 + 1.34 + 0.22$ $= 5.85$
Required value	$= 3.84$ from Table ST11, with 1 degree of freedom.
Conclusion	As the calculated value is greater than the required value we reject the null hypothesis and conclude that the adjustment was effective.

The alternative approach would compare the sample percentage before the adjustment (22.5%) with the sample percentage after the adjustment (11.4%). The overall percentage is 14.33% and 95% confidence limits for the difference between the population percentages are:

$$(p_1 - p_2) \pm 1.96\sqrt{\{p(100-p)[1/n_1) + (1/n_2)]\}}$$
$$= (22.5 - 11.4) \pm 1.96\sqrt{\{14.33\ (85.67)[(1/80) + (1/220)]\}}$$
$$= 11.1 \pm 9.0$$
$$= 2.1\% \ to \ 20.1\%$$

As this confidence interval does not include zero we can conclude that the population percentages differ. In other words, we conclude that the adjustment did reduce the percentage hazing.

(e) In this situation it is appropriate to use a one-sided significance test, although I have used a two-sided test in parts (a), (b) and (d). When George Banks assesses the furnace performance every hour, he is looking for evidence of deterioration. He is expecting a change in one direction. Only if the sample percentage were greater than 12% would he consider whether or not the evidence of deterioration was sufficiently strong to merit stopping the process. To carry out a one-sided test with a 5% significance level we would use the 90% column of the appropriate table.

Solution to Exercise 6.8

(a) The two sample t-test, Wilcoxon's rank sum test and Tukey's quick test could be used.

(b)   Centre spot: mean $= 25.2$ kg   SD $= 2.6192$ kg
      Diagonal stroke: mean $= 28.7$ kg   SD $= 2.4663$ kg
      Combined SD $= 2.5439$ kg

The two-sample t-test:

Null hypothesis	The population mean strength is the same with both methods of application
Calculated value	$= (\bar{x}_1 - \bar{x}_2)/\{s\sqrt{[(1/n_1) + (1/n_2)]}\}$ $= (28.7 - 25.2)/\{2.5439\sqrt{[(1/8) + (1/8)]}\}$ $= 2.75$
Required value	$= 2.15$, from Table ST1, with 14 degrees of freedom, and 5% significance.
Conclusion	As the calculated value exceeds the required value we reject the null hypothesis and conclude that the diagonal stroke method gives greater bond strength than does the centre spot method.

Wilcoxon's rank some test:

Null hypothesis	The population median bond strength is the same with both methods of application.
Calculated value	$=$ The larger of the two rank sums $= 91$
Required value	$= 86.5$ from table ST10 with sample sizes of 8 and 8.
Conclusion	As the calculated value is greater than the required value we reject the null hypothesis and conclude that the diagonal stroke method gives greater bond strength than does the centre spot method.

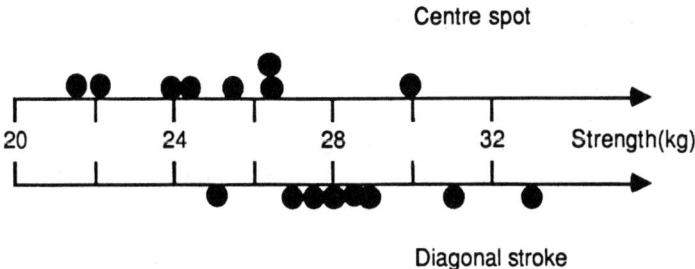

Fig. 6.3 — A belated dot plot.

Tukey's quick test:

The dot plot is helpful when we are carrying out Tukey's quick test. However, it would have been wise to have drawn the dot plot before you did the t-test and the Wilcoxon test.

Null hypothesis	The population mean bond strength is the same with both methods of application.
Calculated value	= The total excedence = 6
Required value	= 7 from your memory.
Conclusion	As the calculated value is less than the required value we cannot reject the null hypothesis. Thus we are unable to conclude that the two methods of application give different bond strength.

(c) The two sample t-test is the most powerful of the three tests but it has the most demanding assumptions. However, there is no evidence in Fig. 6.3 that these assumptions are violated. We can see in the dot plot why the Tukey test has led us to a different conclusion. It is strongly influenced by the 30.0 in the centre spot data. If we exclude this observation, in accordance with Neave's suggestion, the calculated value becomes 11, which exceeds the increased required value of 10.

(d) If you carry out all three significance tests without drawing a dot plot, you should feel thoroughly ashamed and utterly disgusted with your performance. I hope that the guilt instilled by this admonition will ensure that, in future, you draw the dot plot **before** carrying out any analysis.

6.13   DETAILED OBJECTIVES FOR THIS CHAPTER

Now that you have studied this chapter and attempted to relate the new ideas to your existing knowledge, you should be able to do the following.

(1) Explain the meaning of the following terms and use them appropriately in suitable contexts:

(a) Wilcoxon's rank sum test;

(b) Tukey's quick test;
(c) χ^2-test;
(d) contingency table;
(e) qualitative and quantitative data;
(f) observed frequencies and expected frequencies;
(g) confidence limits for a population percentage;
(h) confidence limits for the difference between two population percentages;
(i) significance level or level of significance;
(j) one-sided and two-sided significance tests.

(2) Carry out a Wilcoxon's rank sum test.
(3) Carry out a Tukey's quick test, and explain how allowance can be made for the presence of a suspected outlier.
(4) Carry out a χ^2-test on a contingency table, or two-way frequency table.
(5) Calculate confidence limits for a population percentage.
(6) Calculate confidence limits for the difference between two population percentages. Explain how these confidence limits can be used to reach the conclusion that would have been given by a χ^2-test.
(7) Calculate the size of sample needed to estimate a population percentage with a specified precision.
(8) Explain how significance levels are related to confidence levels and to the risks involved in significance testing.
(9) Explain the difference between a one-sided and two-sided significance test.
(10) Report the results of a significance test with such clarity that you will not run the risk of deceiving the reader of your report.

6.14 SELF-TEST

Dr Ball is investigating the behaviour of two species of insect which attack a variety of maize grown in Brazuela. He has reason to believe that members of both species tend to fly either at a rather high level or at a much lower level. He therefore sets up traps at 1 m and 15 m above ground level. The numbers of insects of each species caught at each level are recorded in Table 6.19 and subdivided by sex in Table 6.20. A random sample of the captured insects were examined and measured. A summary of the lengths of these selected insects is given in Table 6.21.

Table 6.19 — Number of insects captured
(both sexes)

	Species	
	A	B
Height		
1 m	16	19
15 m	14	45

Table 6.20 — Number of insects captured

	Species			
	A		B	
	Male	Female	Male	Female
Height				
1 m	10	6	13	6
15 m	2	12	16	29

Table 6.21 — Lengths (mm) of males and females in both species

	Species			
	A		B	
	Male	Female	Male	Female
Number	12	18	39	35
Mean	28.0	26.3	25.8	22.4
SD	30.6	24.5	19.4	18.6

(1) Which type of significance test would you use to answer the question 'Does one species tend to fly at a greater height than the other?'
   (a) a χ^2-test;
   (b) a two-sample t-test;
   (c) Wilcoxon's rank sum test;
   (d) a correlation test.

(2) Which type of significance test could you use to answer the question 'Do the high flying insects differ in length from the low flying insects?'
   (a) a χ^2-test;
   (b) Wilcoxon's rank sum test;
   (c) Wilcoxon's matched pairs test;
   (d) a correlation test.

(3) How many insects of species A would Dr Ball need to capture in order to estimate the percentage of low flyers in this species to within $\pm$ 0.5%?
   (a) 40
   (b) 400
   (c) 4000
   (d) 40 000

(4) Which of the following are 95% confidence limits for the difference between the percentages of low flyers in species A and the percentage of low flyers in species B?
   (a) 2.6% to 44.6%
   (b) 12.9% to 34.3%
   (c) 14.5% to 32.7%
   (d) 19.8% to 27.4%

(5) Use the correct answer to question (4) to predict the conclusion that would be reached by doing a χ^2-test on the data in Table 6.19. The conclusion would be:

(a) that insects of species A were more likely to fly low than insects of species B;

(b) that insects of species B were more likely to fly low than insects of species A;

(c) that low flying was equally likely with either species;

(d) that low flying was less common than high flying with both species.

(6) Dr Ball wrote in his report, 'In species A the males were more likely to fly lower than the females ($\chi^2 = 7.23$, df $= 1$, $p < 0.01$). However, this cannot be explained by any difference in length between males and females ($t = 0.17$, df $= 28$, $p > 0.05$)'. What do these two sentences imply?

(a) In both significance tests the calculated values were less than the required value from the 95% confidence column.

(b) In both significance tests the calculated values were greater than the required value from the 95% confidence column.

(c) In only one of the two significance tests was the calculated value less than required value from the 95% confidence column.

(d) None of the above.

(7) What evidence is there in Table 6.21 that it would not be valid to use a two-sample t-test with the data in that table?

(a) There is strong evidence that males are more variable in length than females.

(b) There is strong evidence that the lengths contain an outlier.

(c) There is strong evidence that variation in length has a skewed distribution.

(d) None of the above.

(8) What evidence is there in Table 6.20 that it would not be valid to do a two-sample t-test with the data in that table?

(a) There is strong evidence that species A is less numerous than species B.

(b) There is strong evidence that females are more numerous than males.

(c) There is strong evidence that variation in length has a skewed distribution.

(d) None of the above.

(9) Why would it be unwise to use Tukey's quick test to compare the two species using the data in Table 6.19?

(a) Because there are considerably more insects from species B than from species A.

(b) Because there are considerably more insects flying at the higher level than at the lower level.

(c) Because the flying heights do not have a normal distribution.

(d) None of the above.

(10) Consider the significance test referred to in question (2). Should it be a one-sided or a two-sided tests?

(a) The test should be one sided, because we know in advance that 1 m is less than 15 m.

(b) The test should be two sided, because we could continue with the test whichever sample mean was the larger.

(c) The test should be one sided, because all non-parametric tests are one sided.

(d) The test should be two-sided, but it would be valid to do a one-sided test, provided that the report clearly stated that the test was two sided.

6.15 ANSWERS TO SELF-TEST QUESTIONS

 (1) (a)
 (2) (b)
 (3) (d)
 (4) (a)
 (5) (a)
 (6) (c)
 (7) (c)
 (8) (d)
 (9) (d)
(10) (b)

7

Detecting sudden changes

7.1 INTRODUCTION

As this is the final chapter, I should like to remind you of what we have covered in the previous six. You may recall that Chapters 1 and 2 dealt with confidence intervals for population means. Chapter 3 focused on the relationships between variables with emphasis on fitting the best straight line and assessing the accuracy of predictions. Thus, after studying the first three chapters, you should have assimilated several very useful techniques for data analysis.

Chapters 4, 5 and 6 were concerned with significance testing. Some of the significance tests discussed were simply more formal methods of doing what we had achieved earlier using confidence intervals. In these three chapters we discussed in more depth the assumptions underlying the techniques introduced earlier. Thus Chapters 4, 5 and 6 should have helped you to consolidate what you learned in the first three chapters. However, you may have felt that your progress was not so rapid as it had been at the outset.

In this final chapter I want to introduce a new technique, called **cusum analysis**. It is quite remarkable. In essence it is very simple and yet it is extremely useful. I am confident that many readers will be able to apply this technique immediately to data they already have. The aims of this chapter are as follows:

(1) The primary aim of this chapter is to equip you to use the cusum post-mortem technique. This will help you, while examining past data, to decide **whether** and **when** changes occurred.
(2) An additional aim is to show you how the cusum post-mortem technique can be adapted to detect changes in process variability.
(3) A third aim, of secondary importance, is to put the cusum technique in a wider context by considering the use of cusums in quality assurance.

7.2 WHEN DID THE IMPURITY INCREASE?

A particular pigment is manufactured by Textile Chemicals Ltd., using a batch process. This pigment has ben sold in various quantities to a variety of customers throughout Europe and the Middle East for many years. Recently several complaints have been received from customers concerning what they considered to be an unacceptably high level of a certain impurity in the pigment. The research and development manager of Textile Chemicals has been asked to investigate the problem and, with the cooperation of the plant manager, he has extracted from the plant records the analytical results on the last 50 batches. These include determinations of yield and impurity, which are given in Table 7.1.

Table 7.1 — Yield and impurity of 50 batches of pigment

Batch number	Yield	Impurity	Batch number	Yield	Impurity
1	88	5.5	26	84	2.2
2	91	8.0	27	87	5.7
3	89	5.2	28	91	1.9
4	92	4.5	29	84	4.0
5	91	7.4	30	90	2.0
6	87	6.3	31	94	3.3
7	90	7.0	32	93	5.6
8	91	7.0	33	90	5.6
9	93	6.7	34	93	3.9
10	91	5.7	35	97	6.0
11	92	7.3	36	91	3.8
12	90	5.3	37	85	2.3
13	89	3.5	38	87	4.1
14	95	2.4	39	90	4.9
15	96	3.8	40	86	5.2
16	94	4.0	41	93	2.6
17	96	4.8	42	90	5.8
18	94	3.0	43	83	7.1
19	95	2.9	44	88	6.2
20	92	1.9	45	92	3.5
21	95	6.5	46	91	5.4
22	86	3.7	47	85	4.7
23	83	1.1	48	93	2.1
24	92	3.4	49	86	3.5
25	94	4.1	50	92	5.7

Clearly the yield and the impurity vary from batch to batch. 'Why do these fluctuations occur?' you might ask. No doubt the plant manager and the research and development manager could suggest many reasons why the impurity and the yield are not constant, but they certainly could not pinpoint the exact cause of every increase or decrease. Thus it would be impossible to produce a complete explanation for the variation. However, we may be able to find a partial explanation if we examine the data carefully and ask searching questions. 'Is there any pattern in the variation or is it simply random?': this is an important question for, if we can find any regularity in the fluctuations, we might be able to deduce the cause of this regularity. Thus we might gain a better understanding of the process. However, in our search for patterns, we must be careful not to overreact to a mere hint of regularity, which might simply be due to chance.

Before we start hunting for patterns it is highly desirable that we have a clear idea of what we are looking for. If there are any patterns in the data, what form might they take? I would like to suggest that there are two possibilities:

(a) There could be an upward or a downward **trend**. Any such trend might persist throughout the data, or it might be present for only a short period.
(b) There could be a sudden increase or decrease. This could be described as a **step** change. Obviously there might be several step changes in such a large set of data.

You will recall that trends were dealt with in Chapter 3. We shall now examine a technique known as **cusum analysis**, which can help us to detect step changes. Before we use this technique, however, I would like you to examine the data carefully. The impurity data for the 50 batches are displayed in Fig. 7.1.

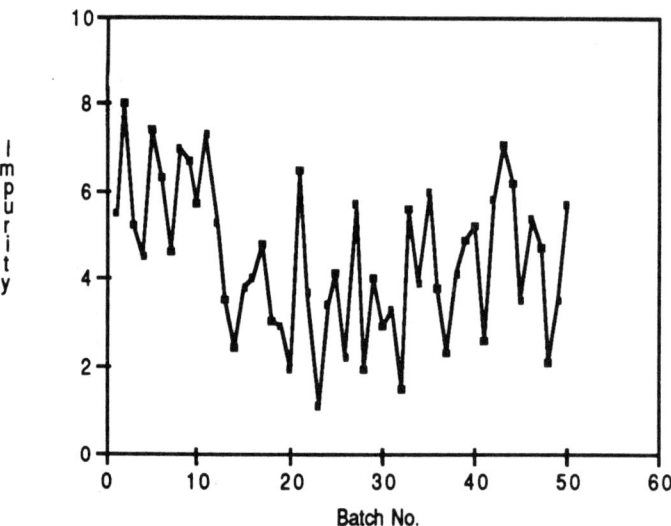

Fig. 7.1 — Impurity in 50 batches of pigment.

Exercise 7.1
(a) Examine Fig. 7.1. While doing so, try to ignore the wild fluctuations from batch to batch. These fluctuations will tend to conceal any long-term changes which

might have occurred. Do you consider that there were any gradual or step changes in impurity during the period when these 50 batches were being produced? The plant manager would like to know, **what type** of changes occurred and **when**?

(b) Now examine Fig. 7.2, which is based on the same data as Fig. 7.1. Do you wish to revise your answer to part (a)?

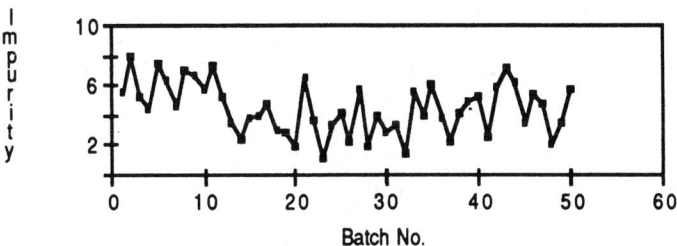

Fig. 7.2 — The data of Fig. 7.1 on a different scale.

(c) Now examine Fig. 7.3, which is also based on the impurity data portrayed in

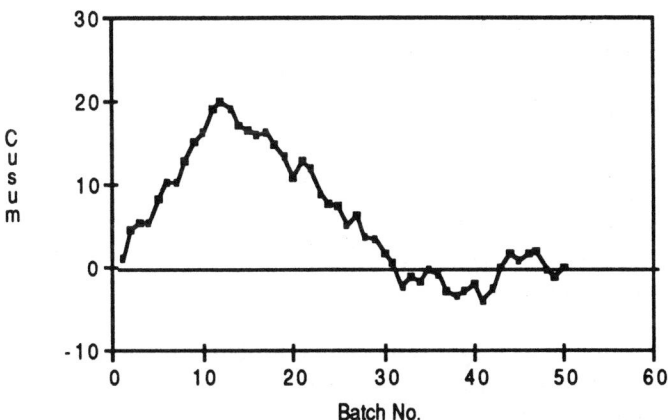

Fig. 7.3 — A cusum plot of the impurity data.

Fig. 7.1 and 7.2, but with a simple modification. This modification has smoothed out the wild fluctuations to reveal a simple pattern. I am sure you can see **changes in slope** in Fig. 7.3. At what points do these changes in slope occur?

Let us now examine how the cusum plot is produced and how it can be interpreted. Both the production and the interpretation are remarkably simple. To calculate the cusum values we subtract a target value from each observation and then calculate a running total, or cumulative total, of the deviations from the target. The calculations are set out in Table 7.2.

Exercise 7.2
In this exercise I want you to consider the calculations in Table 7.2 and how they relate to the shape of the cusum plot in Fig. 7.3. Please insert the missing words, or choose the correct alternative words, in the following:

(a) The target used in the calculation of the deviations was _____ . This is equal to the _____ impurity for all 50 batches.
(b) Because we have used the mean as the target value the total of the 'deviations from target' will be _____ and the final value in the cusum column will be _____ .
(c) The cusum plot has a positive slope for batches 1 to 12. During this period the deviations were positive/negative, because the impurities were above/below average.
(d) The cusum plot has a positive/negative slope for batches 13 to 32. During this period most of the deviations were positive/negative, because most of the impurities were above/below average.
(e) For batches 33 to 50 the slope of the cusum plot is approximately zero. During this period there is a mixture of _____ and _____ deviations, because the mean impurity for batches 33 to 50 is approximately _____ to the target value of 4.45.

Cusum analysis is very simple, but very useful. Furthermore, it can be used in various ways. We have just carried out what could best be described as a **cusum post-mortem analysis**. All the data were available before we started the analysis and we used the **mean** of the whole set of data as the target value. In quality control applications of cusums, the target value would probably be dictated by a specification. Regardless of how or why we are using a cusum analysis, we will be interested in the **slope** of the cusum graph.

When we examine a cusum plot we look for changes in slope. These are indicative of step changes in the mean level of the measured variable. Our cusum post-mortem analysis has led us to believe that the mean impurity changed twice during the period when these 50 batches were being produced. For the first 12 batches the mean impurity was rather high, 6.125. For batches 13 to 32 the mean impurity was much lower, 3.330, and for batches 33 to 50 the mean impurity was higher, 4.4578, but not so high as it had been in the early batches. The step changes in mean that we have detected are illustrated in Fig. 7.4. Such diagrams are often referred to as Manhattan diagrams, because of their resemblance to the skyline of that part of New York.

Table 7.2 — Calculation of the cusum

Batch number	Impurity	Deviation from target value	Cusum
1	5.5	1.05	1.05
2	8.0	3.55	4.60
3	5.2	0.75	5.35
4	4.5	0.05	5.40
5	7.4	2.95	8.35
6	6.3	1.85	10.20
7	4.6	0.15	10.35
8	7.0	2.55	12.90
9	6.7	2.25	15.15
10	5.7	1.25	16.40
11	7.3	2.85	19.25
12	5.3	0.85	20.10
13	3.5	−0.95	19.15
14	2.4	−2.05	17.10
15	3.8	−0.65	16.45
16	4.0	−0.45	16.00
17	4.8	0.35	16.35
18	3.0	−1.45	14.90
19	2.9	−1.55	13.35
20	1.9	−2.55	10.80
21	6.5	2.05	12.85
22	3.7	−0.75	12.10
23	1.1	−3.35	8.75
24	3.4	−1.05	7.70
25	4.1	−0.35	7.35
26	2.2	−2.25	5.10
27	5.7	1.25	6.35
28	1.9	−2.55	3.80
29	4.0	−0.55	3.35
30	2.9	−1.55	1.80
31	3.3	−1.15	0.65
32	1.5	−2.95	−2.30
33	5.6	1.05	−1.15
34	3.9	−0.55	−1.70
35	6.0	1.55	−0.15
36	3.8	−0.65	−0.80
37	2.3	−2.15	−2.95
38	4.1	−0.35	−3.30
39	4.9	0.45	−2.85
40	5.2	0.75	−2.10
41	2.6	−1.85	−3.95
42	5.8	1.35	−2.60
43	7.1	2.65	0.05
44	6.2	1.75	1.80
45	3.5	−0.95	0.85
46	5.4	0.95	1.80
47	4.7	?	?
48	2.1	?	?
49	3.5	?	?
50	5.7	?	?
Total	222.5		
Mean	4.45		

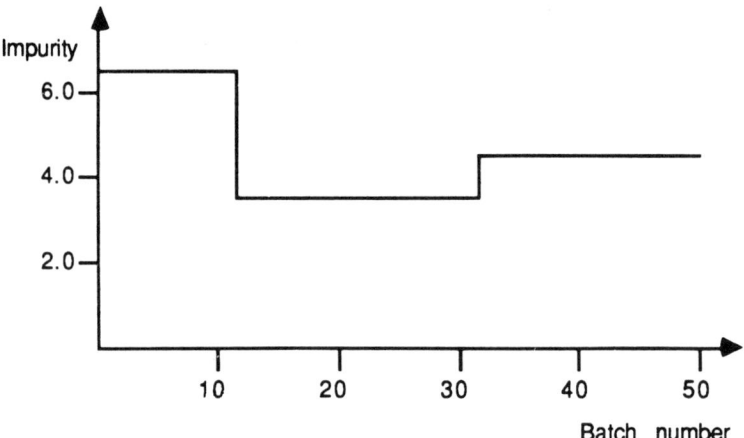

Fig. 7.4 — Changes in mean impurity.

7.3 SIGNIFICANCE TESTING

In the solution to Exercise 7.1, I suggested that there were two changes in slope in the cusum plot. How many changes did you find? I feel sure you found at least one. Perhaps you found the two that I detected. Perhaps you found more than two. If you have good eyesight and a vivid imagination, you may have concluded that there were as many as 49 changes in slope. Obviously we need a more **objective** method for deciding how many slope changes there are in a cusum plot.

The purpose of the analysis is to detect any real changes which may have occurred, but our progress is hampered by the presence of random variation. In essence this is the same problem that we faced in Chapters 4, 5 and 6. You may recall that, in those chapters, we used significance tests to distinguish between real and spurious differences. Let us now use a significance test to assess the changes in slope of the cusum plot. This test will follow the familiar four steps of null hypothesis, calculated value, required value and conclusion.

To obtain the calculated value we take the maximum value from the cusum column in Table 7.2 and divide it by a suitable standard deviation. This standard deviation needs to reflect the batch-to-batch variation in impurity that would be present if there were no real changes in mean impurity. We will use what is known as the localized standard deviation. It cannot be obtained directly from your calculator as it is based on a different formula:

$$\text{localized SD} = \sqrt{\{[(\text{deviation from previous batch})^2]/[2(n-1)]\}}$$
$$= \sqrt{\{212.32/[2(50-1)]\}}$$
$$= 1.4719$$

The 212.32 that was substituted into the formula could be described as the 'sum of squared deviations from previous batches'. It is obtained from the final column of

Table 7.3. Obviously the difference in impurity between two successive batches is likely to reflect batch-to-batch variation. By summing the squares of these differences and dividing by $2(n-1)$, and then taking the square root, we obtain a standard deviation with $n-1$ degrees of freedom.

You may be a little suspicious of the localized standard deviation if you have not met it before. Let us relate it to something with which you are more familiar. Compare the localized standard deviation with the conventional standard deviation in Table 7.4.

Exercise 7.3

(a) Turn to Table 7.3 and calculate the four missing deviations and the four missing squared deviations in the last two columns.

(b) If all 50 batches had the **same** impurity, say 4.45, what would we find in the last two columns of Table 7.3 and what would the localized standard deviation be equal to?

(c) Imagine a hypothetical set of data in which the first 12 batches each had an impurity of 6.0%, batches 13 to 32 each had an impurity of 3.0% and the last 18 batches each had an impurity of 5.0%. What means and standard deviations would these data give in a table like Table 7.4? What would be the localized standard deviation for these hypothetical data?

(d) In Table 7.4 the short-term standard deviations (1.1569, 1.3522 and 1.4318), are all less than the long-term standard deviation (1.7105). Why is this so? (Hint: you may have observed something similar in part (c).)

Now that we have a suitable standard deviation we can carry out a significance test on each of the suspected changes indicated by the cusum plot. We shall test them one by one and we start by finding the largest value in the cusum column of Table 7.2. This is equal to 20.1 (note that if the maximum cusum were negative we would ignore the minus sign). We then divide the maximum cusum by the localised standard deviation to obtain calculated value.

The cusum test

Null hypothesis   There was no change in the population mean impurity during the period when these 50 batches were produced.

Calculated value $=$ (Maximum cusum)/(localized SD)

$=20.1/1.4719$

$=13.66$

Required value $=9.1$ from Table ST12, for a span of 50 observations and 95% confidence, or 5% significance.

Conclusion   As the calculated value is greater than the required value we reject the null hypothesis and conclude that the population mean did change.

The maximum cusum occurred at batch number 12. Thus we have proved,

Table 7.3 — Calculation of the localized SD

Batch number	Impurity	Deviation from previous batch	Squared deviation
1	5.5	—	—
2	8.0	2.5	6.25
3	5.2	−2.8	7.84
4	4.5	−0.7	0.49
5	7.4	2.9	8.41
6	6.3	−1.1	1.21
7	4.6	−1.7	2.89
8	7.0	2.4	5.76
9	6.7	−0.3	0.09
10	5.7	−1.0	1.00
11	7.3	1.6	2.56
12	5.3	−2.0	4.00
13	3.5	−1.8	3.24
14	2.4	−1.1	1.21
15	3.8	1.4	1.96
16	4.0	0.2	0.04
17	4.8	0.8	0.64
18	3.0	−1.8	3.24
19	2.9	−0.1	0.01
20	1.9	−1.0	1.00
21	6.5	4.6	21.16
22	3.7	−2.8	7.84
23	1.1	−2.6	6.76
24	3.4	2.3	5.29
25	4.1	0.7	0.49
26	2.2	−1.9	3.61
27	5.7	3.5	12.25
28	1.9	−3.8	14.44
29	4.0	2.1	4.41
30	2.9	−1.1	1.21
31	3.3	0.4	0.16
32	1.5	−1.8	3.24
33	5.6	4.1	16.81
34	3.9	−1.7	2.89
35	6.0	2.1	4.41
36	3.8	−2.2	4.84
37	2.3	−1.5	2.25
38	4.1	1.8	3.24
39	4.9	0.8	−0.64
40	5.2	0.3	0.09
41	2.6	−2.6	6.76
42	5.8	3.2	10.21
43	7.1	1.3	1.69
44	6.2	−0.9	0.81
45	3.5	−2.7	7.29
46	5.4	1.9	3.61
47	4.7	?	?
48	2.1	?	?
49	3.5	?	?
50	5.7	?	?
Total	222.5	?	212.32

Table 7.4 — Means and SDs for the three groups of batches

Batches	Mean impurity	Standard deviation
1 to 12	6.125	1.1569
13 to 32	3.330	1.3522
33 to 50	4.578	1.4318
1 to 50	4.450	1.7105

beyond reasonable doubt, that the long-term mean impurity decreases at about the time that batch number 12 was produced. Now we can return to the cusum plot to see whether there was a second change in the mean. You may recall that, while examining Fig. 7.3, we noticed a second change in slope at about batch number 32. However, Table 7.2 does not give a large cusum at this point. In fact Table 7.2 is no longer of use to us, as it stands. In order to find and assess additional changes we must either:

(a) produce two new cusum tables, similar to Table 7.2, but treating batches 1 to 13 as one set of data and batches 13 to 50 as a completely separate set of data, or
(b) modify the cusum plot.

If you were using an interactive computer program, it would be very easy to produce two new cusum tables. However, if you were doing the analysis by hand, you would find the second alternative more attractive. We shall modify the cusum plot by drawing on it two straight lines which will join the maximum cusum point to the start and the finish of the plot.

For our second cusum test we shall take the maximum cusum from Fig. 7.5. To

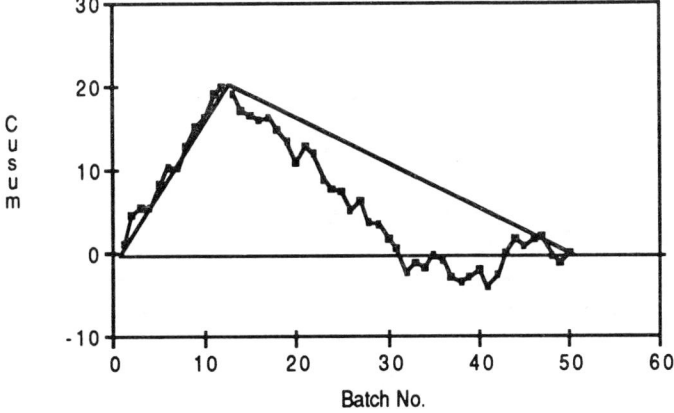

Fig. 7.5 — Modified cusum plot.

obtain it we find the point on the cusum plot which is furthest from the two new lines. This point is at batch number 32 and its distance from the new line is 11.82. Thus, for the second test, our maximum cusum will be 11.82. This will be divided by the localised standard deviation to obtain the calculated value. We could use the 1.4719 that we used in the first cusum test but, to be strictly correct, we should recalculate the localized standard deviation using only batches 13 to 50.

Exercise 7.4
(a) To recalculate the localized standard deviation we can use the squared deviations in the final column of Table 7.3. Add the squared deviations for batches 13 to 50 inclusive and use this total to calcuate the new localized standard deviation.
(b) Carry out a cusum test on the maximum cusum obtained from Fig. 7.5.
(c) How would you now proceed to locate and test a third change in long-term mean impurity?

7.4   DID THE YIELD ALSO CHANGE?

The research and development manager is now convinced that two changes in impurity occurred during the period when these 50 batches were produced. It appears that the mean impurity decreased after batch number 12 and later increased about the time that batch number 32 was manufactured. In order to ascertain the cause or causes of these changes he will discuss the problem with the plant manager and the shift supervisors. However, before he convenes a meeting, he would like to know what was happening to the mean yield during this period. Let us therefore perform a cusum analysis on the yield data in Table 7.1 (Fig. 7.6).

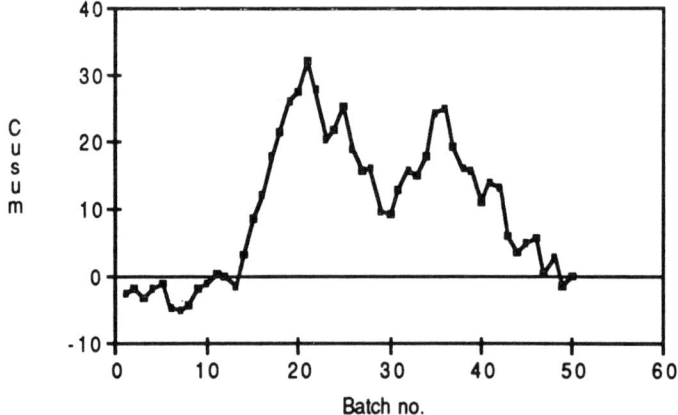

Fig. 7.6 — A cusum plot of the yield data.

What conclusions can we draw from Fig. 7.6? Bearing in mind that the **slope** of the cusum plot is indicative of the mean yield, we look for changes in slope. There is a clear indication of a change in slope at batch number 21. There is also some evidence of change at batch number 10, or is it a little later, perhaps at batch number 13? Carrying out a formal cusum analysis of the yield data reveals two significant changes in mean yield. These are at batches 13 and 21. The mean yield for each group of batches is given in Table 7.5.

Table 7.5 — Changes in mean yield

Batches	Mean yield	Standard deviation
1 to 13	90.31	1.702
14 to 21	94.63	1.302
22 to 50	89.31	3.818
1 to 50	90.42	3.5925

The changes in mean impurity and the changes in mean yield are illustrated in Fig. 7.7. The research and development manager believes that this diagram will form

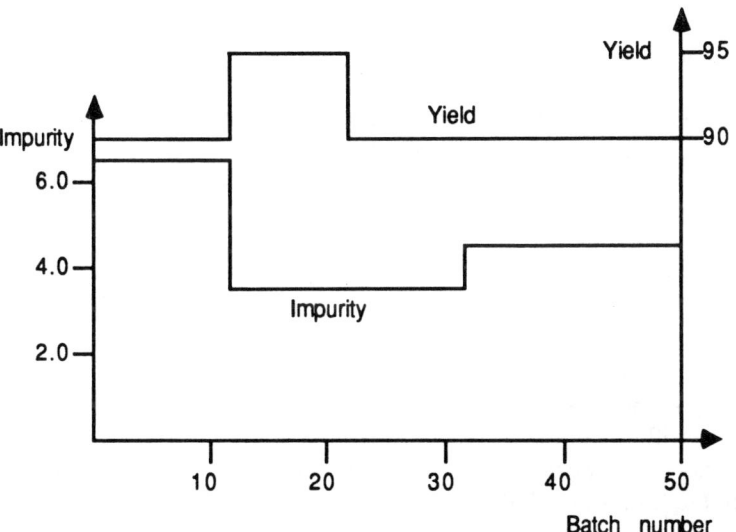

Fig. 7.7 — Changes in yield and impurity.

an excellent focus for discussion. He presents it to the plant manager and two of his shift supervisors. Each of the suspected changes is discussed in turn, and an explanation is put forward by each person attending the meeting.

(a) The supervisor of shift A suggests that the decrease in mean impurity at batch 12 was due to the introduction of the triazone additive. However, he can think of no reason why the impurity should have increased at batch 32. He refuses to believe that there have been any real changs in yield, over and above the normal batch-to-batch fluctuations.

(b) The supervisor of shift B agrees that the decrease in impurity at batch 12 was caused by the introduction of the additive. He is surprised to learn that the mean impurity later increased and wonders whether this could be due to operators' using less than the specified quantity. The operators believe that the introduction of the additive slowed the reaction and caused several batch times to run beyond shift change-over. He is prepared to accept that the mean yield increased about the same time that the impurity decreased. He feels sure that this resulted from the training which the operators received prior to the introduction of the triazone additive. However, he is not prepared to accept that the yield later decreased.

(c) The plant manager is not impressed by Fig. 7.7 nor is he influenced by arguments couched in statistical jargon. He prefers to examine the data in Table 7.1 and to draw his own conclusions about what changes in yield and impurity have occurred. He admits that there is evidence of a short-lived increase in yield between batches 13 and 21. He also believes that there may have been a rather more enduring decrease in impurity between batches 13 and 32. However, his most interesting conclusion is that 'the batch-to-batch variation in yield is much greater in the later batches'.

(d) The research and development manager is rather disturbed by the plant manager's final comment. He wonders whether the process really is more variable than it used to be. Furthermore, he is concerned that any change in process variability might invalidate the cusum analysis which revealed the changes in mean. Thirdly, he is upset because he, himself, had not spotted any change of variability when he examined the data.

Exercise 7.5
The results of a cusum post-mortem analysis can form an excellent basis for the discussion of possible changes. It is quite possible, of course, that the participants in the discussion will not agree on exactly what changes occurred and when. Furthermore, they may not agree on the causes of any changes that did occur. What are your thoughts on the four views that have been expressed?

(a) Comment on the first supervisor's opinion that 'there were no real changes in yield, over and above the normal batch-to-batch fluctuations'.

(b) Comment on the second supervisor's refusal to accept that 'the yield later decreased'. Bear in mind that he would have considerable knowledge of the performance of his own shift team, but much less knowledge of the other two.

(c) Comment on the plant manager's conclusion that 'the batch-to-batch variation in yield is much greater in later batches'.

(d) Comment on the research and development manager's concern that '. . . any change in process variability might invalidate the cusum analysis which revealed the changes in mean'.

7.5   HAS THE PROCESS VARIABILITY INCREASED?

With most chemical production processes we would like to see the impurities in the end product as low as possible. Furthermore, we would like to see the yield as high as possible. It is obviously useful, therefore, to employ the cusum post-mortem technique to seek out long-term changes in mean impurity and mean yield. If we discover causes for these changes we might improve our understanding of the process.

Continuous improvement of systems and processes is a central theme of the philosophy of total quality management. Improvement can be achieved by the shifting of mean levels in the appropriate direction, but it can also be achieved by reducing process variability. Thus changes in variability may be just as important as changes in mean. Fortunately our cusum post-mortem technique can easily be adapted to detect an increase or decrease in variability.

To detect changes in variability we first calculate the difference between successive observations, then we carry out a cusum analysis on these differences. The calculations are set out in Table 7.6.

The absolute differences in Table 7.6 are obtained by subtracting each yield from the next and deleting the minus sign if one arises. Clearly these differences tell us something about the batch-to-batch variation in yield. If there were no variation, all the absolute differences would be equal to zero. If the batch-to-batch variation increased, the differences would be larger.

Examine the column of differences. Do you get the impression that they increase in size after batch 21? If there were such an increase it would support the plant manager's assertion that 'the batch-to-batch variation in yield is much greater in later batches'.

To see whether there is a sudden increase in the differences we carry out a cusum post-mortem analysis using the mean difference, 3.837, as the target value. You will see that the maximum cusum in the final column is 31.740. This proves to be statistically significant when divided by the localized standard deviation and compared with the required value from Table ST12. Thus we conclude that the batch-to-batch variation in yield **did** increase after batch 21.

What is the cause of this increase in variability? If we cannot identify why the increase occurred how can we advise the plant manager how to restore the process to normal performance? Fortunately, an explanation is easily found if we examine in more detail the plant records. We can see in Table 7.7 exactly which batches were made by which shift teams.

If we examine the data from batch 21 onwards we can see that many of the low yield batches were made by shift C. Perhaps this will be seen more clearly if we separate the data for the three shift teams. This has been done in Table 7.8.

This summary of the data reveals that shift team C is responsible for the significant changes in yield that were indicated by the cusum analyses. The decrease

Table 7.6 — A cusum on differences

Batch number	Yield	Absolute difference	Deviation from target	Cusum
1	88	—	—	—
2	91	3	−0.837	−0.837
3	89	2	−1.837	−2.674
4	92	3	−0.837	−3.511
5	91	1	−2.837	−6.348
6	87	4	0.163	−6.185
7	90	3	−0.837	−7.022
8	91	1	−2.837	−9.859
9	93	2	−1.837	−11.696
10	91	2	−1.837	−13.533
11	92	1	−2.837	−16.370
12	90	2	−1.837	−18.207
13	89	1	−2.837	−21.044
14	95	6	2.163	−18.881
15	96	1	−2.837	−21.718
16	94	2	−1.837	−23.555
17	96	2	−1.837	−25.392
18	94	2	−1.837	−27.329
19	95	1	−2.837	−27.229
20	92	3	−0.837	−30.903
21	95	3	−0.837	−31.740
22	86	9	5.163	−26.577
23	83	3	−0.837	−27.414
24	92	9	5.163	−22.251
25	94	2	−1.837	−24.088
26	84	10	6.163	−17.925
27	87	3	−0.837	−18.762
28	91	4	0.163	−18.599
29	84	7	3.163	−15.436
30	90	6	2.163	−13.273
31	94	4	0.163	−13.110
32	93	1	−2.827	−15.947
33	90	3	−0.837	−16.784
34	93	3	−0.837	−17.621
35	97	4	0.163	−17.458
36	91	6	2.163	−15.295
37	85	6	2.163	−13.132
38	87	2	−1.837	−14.969
39	90	3	−0.837	−15.806
40	86	4	0.163	15.643
41	93	7	3.163	−12.480
42	90	3	−0.837	−13.317
43	83	7	3.163	−10.154
44	88	5	1.163	−8.991
45	92	4	0.163	−8.828
46	91	1	−2.837	−11.665
47	85	6	2.163	−9.502
48	93	8	4.163	−5.339
49	86	7	3.163	−2.176
50	92	6	2.163	−0.012
Total	4521	188	−0.013	—
Mean	90.42	3.837	−0.000	—

Table 7.7 — Which shift team made the poor batches?

Batch number	1	2	3	4	5	6	7	8	9	10	11	12	13
Shift team	B	C	A	B	C	A	A	B	C	B	C	A	B
Yield	88	91	89	92	91	87	90	91	93	91	92	90	89
Batch number	14	15	16	17	18	19	20	21	22	23	24	25	26
Shift team	C	B	C	A	B	C	C	A	B	C	A	B	C
Yield	95	96	94	96	94	95	92	95	86	83	92	94	84
Batch number	27	28	29	30	31	32	33	34	35	36	37	38	39
Shift team	A	B	C	C	A	B	C	A	A	B	C	A	B
Yield	87	91	84	90	94	93	90	93	97	91	85	87	90
Batch number	40	41	42	43	44	45	46	47	48	49	50		
Shift team	C	A	B	C	A	B	B	C	A	C	A		
Yield	86	93	90	83	88	92	91	85	93	86	92		

Table 7.8 — A summary of Table 7.7

Team		Batches 1 to 21			Batches 22 to 50	
	n	Mean	SD	*n*	Mean	SD
A	6	91.2	3.45	10	91.6	3.27
B	7	91.6	2.76	9	90.8	2.26
C	8	92.9	1.64	10	85.6	2.55

in mean yield from 92.9 to 85.6, for this team, has contributed considerably to one of the changes in mean shown in Fig. 7.7 and has inflated the variability in the later batches that we saw in Table 7.5.

7.6 CUSUMS IN QUALITY ASSURANCE

This chapter has been devoted entirely to the cusum post-mortem technique. Clearly it is a simple technique, although its execution is rather tedious if you do not have access to a computer program. Despite its simplicity, however, the cusum post mortem is a very useful technique.

Before we close this discussion, I must point out that cusums have many other applications, in addition to post mortems. Perhaps their most widespread use is in quality assurance and/or statistical process control, but it would not be appropriate to include a full treatment of this application in this book. A later volume in the series will be devoted to the use of statistical techniques in quality problems. However, I feel that I should illustrate how a cusum analysis in quality assurance would differ from a post-morten analysis.

In the cusum post-mortem analysis we started with a complete set of data and we first calculated the mean to use as the target value. Then we produced the cusum plot, looked for changes in slope and checked their statistical significance. In a quality assurance application of cusums we would start with no data at all. The target value might be imposed on us by a product specification or it might be a quality level that we deemed appropriate.

We would receive the data over a period of time, with quality determinations arriving at intervals, probably regular intervals. Each new data item would give us an additional point on the cusum plot. Thus the plot would start as one point and would grow and grow .... As each new point was added to the graph we would ask the question, 'Did the **slope** of the cusum plot change at any time?'

Consider, for example, Figs. 7.8, 7.9 and 7.10, which show how the cusum plot

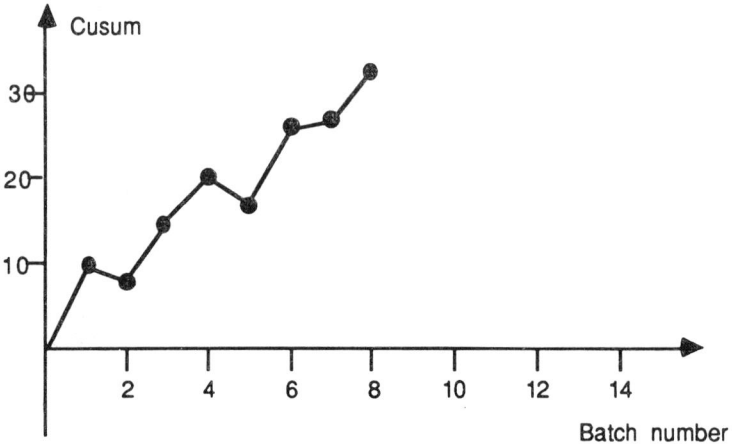

Fig. 7.8 — The cusum graph after plotting eight points.

might appear after we have plotted 8, 9 and then 10 points. In Fig. 7.8 there is evidence that the mean quality has been above the target value since the plot was started. By the time we have plotted the ninth point in Fig. 7.9 we might begin to suspect that the mean impurity has decreased. After the tenth point has been plotted we might feel that our initial suspicion has been confirmed, or we might feel that an eleventh point is needed before we can be sure.

After studying Figs. 7.8, 7.9 and 7.10 you should appreciate the difference between this quality assurance application of cusums and the post-mortem analysis. I hope that you will have learned the following from these three diagrams:

(a) If a change in mean occurs, we are unlikely to detect the change **immediately**.

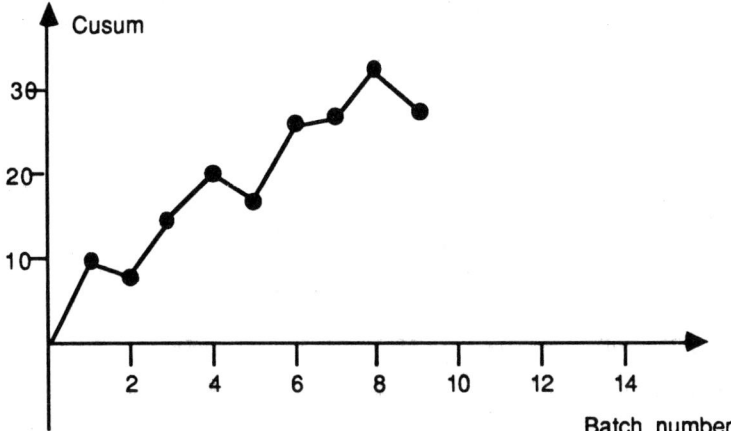

Fig. 7.9 — The cusum graph after plotting nine points.

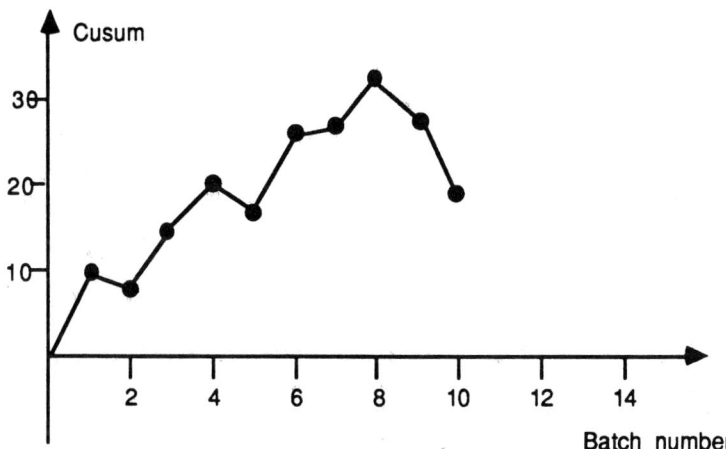

Fig. 7.10 — The cusum graph after plotting ten points.

(b) If a change in mean occurs, we will probably detect it **sooner or later**, but a large change in mean will probably be detected more quickly than a small change.

(c) We need an objective method for deciding whether or not a change in mean has occurred.

In practice the decision making is formalized by the use of what is known as a **V-mask**. The V-mask consists of a transparent sheet on which are drawn three lines and a point. The mask is laid on the cusum plot so that this point coincides with that most recently plotted. The placing of the V-mask is illustrated in Fig. 7.11.

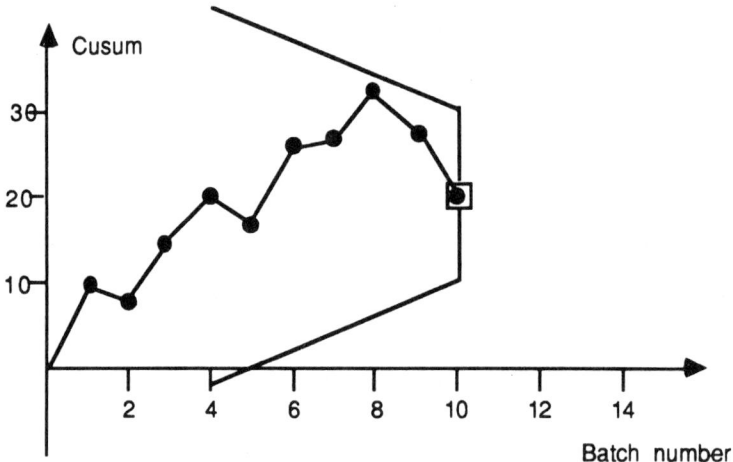

Fig. 7.11 — A V-mask laid on the cusum plot.

When the mask is placed on the graph we ask 'Is any part of the cusum plot outside the V?' In Fig. 7.11 the answer is 'no', so we conclude that the mean impurity has not changed. However, we can see that, if the cusum continues with a negative slope as it has for the last three observations, the answer will soon be 'yes'. When we find that one or more points lie **outside** the V-mask we conclude that the mean impurity has changed. Our best estimate of **when** the mean changed is given by the earliest point which lies outside the V.

Exercise 7.6
Whenever we use a V-mask to make a decision we cannot be **certain** that we have made the corrrect decision.

(a) Describe the two possible errors that we might make each time we place the V-mask on the cusum plot.
(b) Suppose we widened the V-mask by making the vertical line longer. How would this affect the chance of making correct decisions?

7.7 A SUMMARY OF THE IMPORTANT POINTS

In many ways this final chapter is the simplest of the seven chapters in this introductory text. The whole chapter has been devoted to one technique, which is inherently simple but surprisingly useful.

(1) **Cusum analysis** is a technique that can be used to detect sudden changes, i.e. step changes, in a set of data.

(2) In any cusum analysis we subtract a **target value** from each observation; then we calculate a cumulative total, or cumulative sum, of the deviations from target.

(3) A graph of the cumulative sum, or cusum, against the observation number is known as a **cusum plot** or cusum graph.

(4) For any period in which the slope of the cusum plot is negative, the mean of the observations is below the target value. In contrast, a positive slope is indicative of a mean above target.

(5) A **change in slope** on the cusum plot indicates a sudden change in mean.

(6) In a **cusum post-mortem analysis** we use the overall mean of the data as the target value.

(7) In a **quality assurance** application of the cusum technique the target value would be imposed by a specification or by some preselected quality level.

(8) In a cusum post-mortem analysis we **test** each suspected change in mean by calculating (maximum cusum)/(localised SD) and comparing this with a required value from Table ST12.

(9) The cusum post-mortem technique can also be used for detecting sudden changes in process **variability**. We adapt the technique for this purpose by first calculating successive differences; then we continue in the usual way.

(10) In a quality assurance application of cusum analysis we use a **V-mask** to help us decide whether or not a change in mean has occurred. This application will be discussed in more detail in a later volume in this series.

7.8   ADDITIONAL EXERCISES

Exercise 7.7

Arthur Robson is employed by Lubo Chemicals as a sales representative. He travels considerable distances by car, visiting manufacturers of lubricating oils. Being very interested in the performance of car engines, he faithfully records the mileage shown on his mileometer whenever he stops to buy petrol. His log book for 1989 contains the entries shown in Table 7.9.

Table 7.9 — Mileage recorded at each refuelling

5186	5321	5444	5578	5719	5846	5972	6104
6224	6346	6487	6616	6737	6885	7022	7154
7290	7410	7513	7629	7760	7864	7987	8102
8205	8308	8430	8545	8664	8784	8895	9026
9152	9286	9403	9524	9634	9748	9870	9983

Throughout this period Arthur Robson has consistently followed the policy of buying exactly four gallons of petrol at the first filling station he passes after his fuel gauge has indicated less than half full. How would you analyse the data in Table 7.9

with the objective of advising Robson as to whether or not his petrol consumption in miles per gallon changed during this period?

Exercise 7.8
(a) For each mileage recorded in Table 7.9 except the first (5186), calculate the mileage travelled since the last purchase.
(b) Carry out a cusum post-mortem analysis on the 39 mileages you calculated in part (a).

Exercise 7.9
The cusum post-mortem technique is very useful, but it is not above criticism. For example, the chance of detecting a relatively short-lived shift in mean will depend on whether it occurs in the middle or at the end of your data. Consider the two sets of data shown in Tables 7.10 and 7.11.

Table 7.10 — Hypothetical data: A

Time	1	2	3	4	5	6	7	8	9	10	11	12
Impurity	3	3	3	3	3	3	3	6	6	6	6	3

Table 7.11 — Hypothetical data: B

Time	1	2	3	4	5	6	7	8	9	10	11	12
Impurity	3	3	3	3	6	6	6	6	3	3	3	3

In both sets of data the impurity is equal to 3 in every batch except for four deviant batches which have greater impurity. You might expect that both sets of data would give the same maximum cusum and the same localized standard deviation.

(a) For data set A calculate the mean impurity, the maximum cusum and the localized standard deviation.
(b) Repeat part (a) for data set B.
(c) If you plotted the two cusum graphs how would their slopes compare?

7.9 WORKED SOLUTIONS

Solution to Exercise 7.1
(a) From Fig. 7.1, I get the impression that the mean level of impurity is lower in the later batches than in the first 11 or 12. Perhaps there was a **sudden** decrease in mean impurity about the time that batches 12, 13 and 14 were produced. I also get the impression that there was a **gradual** increase in impurity from batch 20 through to batch 50. I would not be surprised, however, if you reached very different conclusions about the long-term changes in impurity.

(b) You may find Fig. 7.2 more helpful than Fig. 7.1. Both represent the impurity data in exactly the same format but a different scale was used on the y-axes. Do you reach the same conclusions that you drew in part (a)?

(c) Fig. 7.3 is very different from the two preceding diagrams. I am sure that you, and I, and all other readers, would agree that the slope of the cusum plot changes at batch 12 and at batch 32. 'What do these changes in slope tell us?' you may wonder. The production and interpretation of cusum plots will now be discussed.

Solution to Exercise 7.2
(a) 4.45; mean.
(b) zero; zero.
(c) positive; above.
(d) negative, negative, below.
(e) negative, positive; equal.

Solution to Exercise 7.3
(a)   **Table 7.12** — Solution to Exercise 7.3(a)

Batch	Deviations from previous batch	Squared deviation
47	0.7	0.49
48	−2.6	6.76
49	1.4	1.96
50	2.2	4.84
Total	0.0	212.32

(b) Every number in the last two columns would be equal to 0.0, the sum of squared deviations would be equal to zero and the localized standard deviation would be zero.

   Thus the localized standard deviation would be a true reflection of the batch-to-batch variation, which is non-existent. The conventional standard deviation of the 50 impurities would also be zero, so it also reflects the batch-to-batch variation.

(c)   **Table 7.13** — Solution to Exercise 7.3(c)

Batches	Mean	SD
1 to 12	6.0	0.000
13 to 32	3.0	0.000
33 to 50	5.0	0.000
1 to 50	4.44	1.2480

47 of the 49 deviations would be equal to zero. The other two would be equal to

−3 and 2. Thus the sum of the squared deviations would be equal to 13.0 and the localized standard deviation would be 0.3642.

(d) The short-term standard deviations are influenced only by the inherent variability of the process. The long-term standard deviation includes the additional variability due to the two step changes. The hypothetical data in part (c) illustrate this point more clearly, perhaps, because the hypothetical process does not have any short-term variability.

Solution to Exercise 7.4

(a) Sum of squared deviations = 168.58

   Localized SD = 168.58/[2(38−1)]

   $\qquad\qquad$ = 1.5093

(b) *Null hypothesis* $\quad$ There was no change in the population mean impurity during the period when batches 13 to 38 were produced.

   Calculated value $\quad$ = 11.82/1.5093

   $\qquad\qquad\qquad$ = 7.83

   Required value $\quad$ = 7.74 approximately, from Table ST12 for a span of 38 observations and 5% significance.

   Conclusion $\quad$ As the calculated value is greater than the required value we reject the null hypothesis and conclude that the population mean impurity did change.

(c) To locate a third change in mean we would draw two straight lines so as to join the cusum point for batch 32 to the cusum points for batches 12 and 50. This would give us the three straight lines in Fig. 7.12. We would then find the cusum point which was furthest from these three lines.

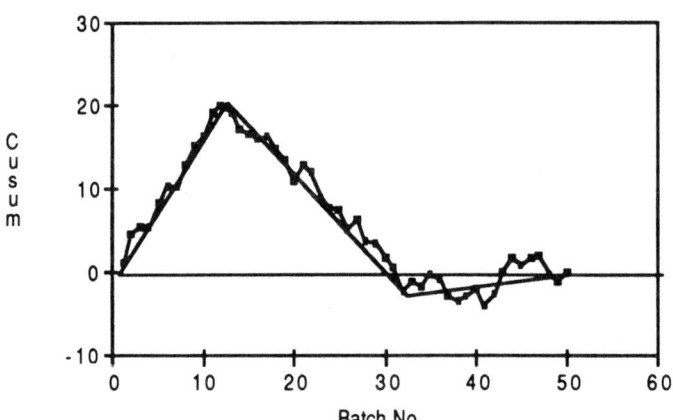

Fig. 7.12 — Seeking a third change in mean.

Solution to Exercise 7.5

(a) We have carried out significance tests which prove beyond reasonable doubt that two step changes in yield occurred. It is always possible that a significance test will lead us to a false conclusion, of course, but the chance of this happening is less than 5% in each case. It is also possible that the shift supervisor has more information than is contained in Table 7.1.

(b) Perhaps the supervisor of shift B is confident that the mean yield did not decrease in later batches because he is considering only those batches produced by his shift team. Could it be that the mean yield decreased in those batches produced by shifts A and C, but remained high for those produced by shift B? If this were the case, what effect would this have on the yield data? We shall discuss this in the next section.

(c) We can see in Table 7.4 that the standard deviation of yield is much greater for batches 22 to 50 than for earlier batches. We could use an SD-test to see whether this increase is statistically significant. Before doing so, however, we would be wise to consider that any increase in process variability might have occurred before batch 21 or after batch 22. We shall see in the next section how the cusum post-mortem analysis can be easily modified to detect changes in variability as well as step changes in mean.

(d) In Chapter 5 we discussed the assumptions underlying the two-sample t-test. You may recall that this test, for comparing two means, is only valid if the population standard deviations are equal. In the cusum test we are also comparing means. We wish to know whether there is a significant difference between the mean yield before a certain time and the mean yield after that time. An assumption underlying this test is that the standard deviation is constant. Can we be confident that the batch-to-batch variation in yield is stable? The plant manager thinks not. The research and development manager is concerned. We shall discuss this problem in the next section.

Solution to Exercise 7.6

(a) The two possible errors that we might make are (i) to conclude that the mean **had** changed, when in fact it had not, and (ii) to conclude that the mean **had not** changed, when in fact it had. We hope that the chance of making the first type of error will be small. We also hope that the chance of making the second type of error will be small. By changing the shape of the V-mask we can **reduce** the chance of the first type of error but, unfortunately, this change will **increase** the chance of the second type's ocurring.

(b) By widening the V-mask we decrease the chance of a point's lying outside the mask. Thus we decrease the chance of concluding that a change in mean has occurred, whether it has or not. Thus we decrease the risk of concluding that a change has occurred, when in fact it has not. At the same time we increase the risk of concluding that the mean has not changed, when in fact it has.

Solution to Exercise 7.7

It would not be useful to carry out a cusum post-mortem analysis on the data as recorded. Obviously each mileage in Table 7.9 is greater than the previous mileage. Thus if we plotted the data against time we would find a positive trend. The best question to ask of such a plot would be 'Has the slope changed?', for the unit of slope is miles per four gallons. To answer this question you could fit the best straight line and then inspect the residuals.

Alternatively, you could subtract each mileage from the next to obtain the miles travelled between purchases. It would then be useful to carry out a cusum post-mortem analysis on these derived data. You are asked to do this in the next exercise.

Solution to Exercise 7.8

(a) 135 123 134 141 127 126 132 120 122 141
    129 121 148 137 132 136 120 103 116 131
    104 123 115 103 103 122 115 119 120 111
    131 126 134 117 121 110 114 122 113

(b) Mean mileage = 123

Maximum cusum = 136.0

Localised standard deviation = 8.867

Calculated value = 15.3

Required value = 8.0

We conclude that the mean milage changed, after purchase number 17, when the car had travelled a total of 7290 miles. From this time onwards petrol consumption was greater (i.e. miles per gallon decreased).

Continuing the cusum post-mortem analysis, we do not find any further significant changes, using a 5% significance level.

Solution to Exercise 7.9

(a) For data set A:

mean impurity = 4.0

maximum cusum = 7.0

localized SD = 0.9045

(b) For data set B:

mean impurity = 4.0

maximum cusum = 4.0

localized SD = 0.9045

(c) You would find exactly the same slopes in the two graphs, but the section with a positive slope would be further to the left in the second graph. Thus **the same change in slope** occurs in both cusum graphs, but the cusum test does not give the same result in both cases.

Obviously it is possible, when two changes occur in the middle of a data set, for them to be missed. For this reason, many users of the cusum post-mortem technique prefer to use a 10% significance level rather than the usual 5%. This increases the chance of finding a real change, but it also increases the chance of a false alarm.

7.10 DETAILED OBJECTIVES FOR THIS CHAPTER

Now that you have studied this chapter and attempted to relate its content to your existing knowledge, you should be able to do the following.

(1) Explain the meaning of the following terms and use them appropriately in suitable contexts:

(a) cusum analysis;

     (b) cusum post-mortem technique;
     (c) target value;
     (d) localized standard deviation;
     (e) cusum plot;
     (f) cusum test;
     (g) Manhattan diagram;
     (h) V-mask.
(2) Draw a **cusum plot**.
(3) Calculate a **localized** standard deviation.
(4) Carry out a **cusum test**.
(5) Draw conclusions concerning step changes in the population mean and illustrate these changes in a Manhattan diagram.
(6) Adapt the cusum post-mortem analysis to seek step changes in **process variability**.
(7) Use a **V-mask** to detect changes in means and explain how the shape of the V-mask affects the chance of drawing false conclusions.

7.11 SELF-TEST

The cusum post-mortem technique is normally used to analyse quite large sets of data. For example, we had data on 50 batches in Table 7.1. However, for the purpose of testing your understanding of this chapter, we shall refer to the much smaller set of data in Table 7.14.

Table 7.14 — Number of items produced in 10 days

Day	1	2	3	4	5	6	7	8	9	10	Mean
Number of items	64	73	69	61	74	69	64	67	58	61	66.0

These data give the number of items produced by a particular process on 10 consecutive working days. The data are referred to in questions (1) to (5) below.

(1) If you carried out a cusum post-morten analysis of the above data what would be the maximum cusum?

     (a) 13.0
     (b) 14.0
     (c) 15.0
     (d) none of the above.

(2) What is the localized standard deviation for these data?
     (a) 4.9
     (b) 5.2
     (c) 5.3
     (d) 5.6

(3) Which of the following would be the conclusion given by a cusum test on the data in Table 7.14?

    (a) that the mean number of items had changed significantly during the ten day period;

    (b) that the mean number of items was significantly greater in the earlier days than in the later days;

    (c) that the localized standard deviation was significantly less than the normal standard deviation;

    (d) that there had been no significant change in mean during the ten day period.

(4) Which of the following would describe the slope of the cusum plot?

    (a) The slope is positive for the early days and negative for the later days.
    (b) The slope changes from negative to positive at day 6.
    (c) The slope changes from positive to negative on day 4.
    (d) The slope is positive from day 2 to day 10.

(5) If you were asked to carry out a cusum analysis on these data, to ascertain whether or not the variability had changed, what would you use as the target value?

    (a) 6.56 approximately;
    (b) 5.90;
    (c) the standard deviation of the ten observations;
    (d) the localized standard deviation.

(6) After finding the first significant change using the cusum post-mortem technique, how would you proceed?

    (a) Discard the data which precede the change and re-analyse the data which follow the change.
    (b) Draw two straight lines on the cusum plot and then find the point furthest from these two lines.
    (c) Discard the data which follow the change and re-analyse the data which precede the change.
    (d) Draw two straight lines on the cusum graph and then find where the cusum plot crosses these two lines.

(7) In a quality assurance application of cusums we use a V-mask in order

    (a) to decide whether or not the target value should be changed;
    (b) to decide whether or not the target value is significantly different from the acceptable standard of quality;
    (c) to decide whether or not the quality has changed from below target to above target;
    (d) to decide whether or not the quality is significantly different from the acceptable standard.

(8) The moisture content of a powder emerging from a continuous belt dryer is measured every 30 minutes and the result used to add a point to a cusum plot. As each new point is added to the graph, how can we decide whether or not the

moisture content is acceptable?

(a) by carrying out a cusum test using the **maximum** cusum and the localized standard deviation;

(b) by carrying out a cusum test using the **new** cusum value and the localized standard deviation;

(c) by answering the question 'Has the cusum changed from negative to positive?';

(d) none of the above.

(9) Which of the following statements is true, concerning a cusum post-mortem analysis carried out on a series of impurity determinations?

(a) A change in the cusum from positive to negative is indicative of an **increase** in the population mean impurity.

(b) A change in the cusum from positive to negative is indicative of a **decrease** in the population mean impurity.

(c) A change in the slope of the cusum graph is indicative of a change in the population mean impurity.

(d) None of the above.

(10) Which of the following statements is true concerning a cusum post-mortem analysis carried out on successive differences calculated from a series of impurity determinations?

(a) A change in the cusum from positive to negative is indicative of an **increase** in the variability of impurity.

(b) A change in the cusum from positive to negative is indicative of an **increase** in the variability of impurity.

(c) A change in the slope of the cusum graph is indicative of a **change** in the variability of impurity.

(d) None of the above.

7.12   ANSWERS TO SELF-TEST QUESTIONS

(1) (b)
(2) (b)
(3) (d)
(4) (a)
(5) (a)
(6) (b)
(7) (d)
(8) (d)
(9) (c)
(10) (c)

Appendix A

To obtain the standard deviation of a set of data we would normally make use of a computer or a pocket calculator. Of course, we must choose a device which has been programmed to carry out the task, then we simply type in the data and out comes the result. For those readers who do not have access to a calculator or computer, and for those who are curious to know what is involved, I will illustrate how a standard deviation can be calculated 'by hand'. Two formulae are offered:

$$\text{standard deviation} = \sqrt{\left\{\left[\Sigma(x - \bar{x})^2\right]\Big/(n-1)\right\}}$$

$$\text{standard deviation} = \sqrt{\left\{\left(\Sigma x^2 - n\bar{x}^2\right)\Big/(n-1)\right\}}$$

Both formulae will give the correct answer provided you do not introduce rounding errors. The second formula is easier to use with most sets of data, but you are likely to get a better understanding of standard deviations by using the first. I will use both formulae to calculate the standard deviation of the second set of data on Exercise 1.2.

Table A.1 — Calculation of SD by the first formula

	Data x	Deviation from mean $(x - \bar{x})$	Squared deviation $(x - \bar{x})^2$
	4	0.5	0.25
	1	− 2.5	6.25
	3	− 0.5	0.25
	2	1.5	2.25
	5	1.5	2.25
	6	2.5	6.25
Total	21	0.0	17.50
Mean	3.5	0.0	—

Standard deviation $= \sqrt{\left\{\left[\Sigma(x-\bar{x})^2\right]\Big/(n-1)\right\}}$

$= \sqrt{\{17.50/5\}}$

$= 1.870\,828\,7$

If we round this result to three decimal places we get 1.871 which agrees with the result given in the worked solutions. Let us now use the second formula to obtain the standard deviation of the same set of data.

Table A.2 — Calculation of SD using the second formula

	Data x	Squared data x^2
	4	16
	1	1
	3	9
	2	4
	5	25
	6	36
Total	21	91
Mean	3.5	—

Standard deviation $= \sqrt{\left\{\left(\Sigma x^2 - n\bar{x}^2\right)\Big/(n-1)\right\}}$

$= \sqrt{\{(91 - 6(3.5)^2)/5\}}$

$= \sqrt{\{17.5/5\}}$

$= 1.870\,828\,7$

We see that both formulae have given the same result. With this set of data the mean, 3.5, was easy to handle. Had the mean of the six observations been 3.333 333, say, we might have been tempted to round this to 3.3, but by doing so we would have introduced errors that would have led to an incorrect value for the standard deviation. You might like to calculate the standard deviation of the following set of data to see what effect this rounding would have:

4, 1, 3, 2, 5, 5

Note that with both formulae we divide by $(n-1)$ before we take the square root. This divisor is often referred to as the 'degrees of freedom' of the standard deviation. In fact both formulae could be written as:

standard deviation $= \sqrt{\{(\text{sum of squares})/(\text{degrees of freedom})\}}$

Clearly the two formulae differ only in the way we calculate the sum of squares, or the 'sum of squared deviations from the mean', to use its full descriptive title.

Perhaps, when you turned to this appendix, you wished to ask the question, 'When calculating a standard deviation, why do we divide by $(n-1)$ rather than n?' Having read the previous paragraph you might reword your question to read 'Why does a standard deviation calculated from n numbers, have $(n-1)$ degrees of freedom?' This question can be answered in many ways. For example, I could point out that we lose a degree of freedom when we subtract the mean from each observation. Thus the 6 numbers in the first column have 6 degrees of freedom, but the 6 deviations in the second column have only 5 degrees of freedom.

A more helpful explanation might emerge if we focus on the reason *why* we calculate the standard deviation of a set of data. Usually we are attempting to estimate the standard deviation of the population from which we have taken a sample. (See section 1.3 for discussion of this objective.) It can be shown, either by repeated experiment or by mathematics, that the two formulae I have offered you give *good* estimates of the population standard deviation. If you were to change the formulae so that you divided by n rather than $(n-1)$, then your revised formulae would give standard deviations which tended to *underestimate* the population standard deviation.

Don't worry about standard deviations. They will become more meaningful as you work your way through the book. Just keep in mind that the standard deviation of a set of data measures the spread, or scatter, or variability in the data. Furthermore, a standard deviation is very easily obtained from a good calculator.

Statistical tables

TABLE ST1 — Confidence intervals and the *t*-test

Degrees of freedom	Confidence level					
	90%	95%	98%	99%	99.8%	99.9%
1	6.31	12.71	31.82	63.66	318.31	636.62
2	2.92	4.30	6.97	9.93	22.33	31.60
3	2.35	3.18	4.54	5.84	10.21	12.92
4	2.13	2.78	3.75	4.60	7.17	8.61
5	2.02	2.57	3.37	4.03	5.89	6.87
6	1.94	2.45	3.14	3.71	5.21	5.96
7	1.90	2.37	3.00	3.50	4.79	5.41
8	1.86	2.31	2.90	3.36	4.50	5.04
9	1.83	2.26	2.82	3.25	4.30	4.78
10	1.81	2.23	2.76	3.17	4.14	4.59
11	1.80	2.20	2.72	3.11	4.03	4.44
12	1.78	2.18	2.68	3.06	3.93	4.32
13	1.77	2.16	2.65	3.01	3.85	4.22
14	1.76	2.15	2.62	2.98	3.79	4.14
15	1.75	2.13	2.60	2.95	3.73	4.07
16	1.75	2.12	2.58	2.92	3.69	4.02
17	1.74	2.11	2.57	2.90	3.65	3.97
18	1.73	2.10	2.55	2.88	3.61	3.92
19	1.73	2.09	2.54	2.86	3.58	3.88
20	1.73	2.09	2.53	2.85	3.55	3.85
25	1.71	2.06	2.49	2.79	3.45	3.73
30	1.70	2.04	2.46	2.75	3.39	3.65
40	1.68	2.02	2.42	2.70	3.31	3.55
60	1.67	2.00	2.39	2.66	3.23	3.46
∞	1.64	1.96	2.33	2.58	3.09	3.29

Confidence interval for the population mean is

$$\bar{x} \pm ts/\sqrt{n}$$

where $\bar{x}$ is the sample mean, t is from the above table, s is the sample standard deviation and n is the sample size (see section 2.3 for more flexible use of this formula.) Calculated values for the various *t*-tests can be obtained using formulae from Table 4.7.

TABLE ST2 — Tolerance intervals

Sample size	Confidence level					
	90%			95%		
	% of items			%of items		
	90%	95%	99%	90%	95%	99%
3	5.847	6.919	8.974	8.380	9.961	—
4	4.166	4.943	6.440	5.369	6.370	8.299
5	3.494	4.152	5.423	4.275	5.079	6.634
6	3.131	3.723	4.870	3.712	4.414	5.775
7	2.902	3.452	4.521	3.309	4.007	5.248
8	2.743	3.364	4.278	3.136	3.732	4.891
9	2.626	3.125	4.098	2.967	3.352	4.631
10	2.535	3.018	3.959	2.839	3.379	4.433
12	2.404	2.863	3.758	2.655	3.162	4.150
14	2.314	2.756	3.618	2.529	3.012	3.955
16	2.246	2.676	3.514	2.437	2.903	3.812
18	2.194	2.614	3.433	2.366	2.819	3.702
20	2.152	2.564	3.368	2.310	2.752	3.615
30	2.025	2.413	3.170	2.140	2.549	3.350
40	1.959	2.334	3.006	2.052	2.445	3.213
50	1.916	2.284	3.001	1.996	2.379	3.126
Infinity	1.645	1.960	2.256	1.645	1.960	2.576

Lower tolerance limit $= \bar{x} - ks$
Upper tolerance limit $= \bar{x} + ks$
where $\bar{x}$ is the sample mean, s is the sample standard deviation, k is taken form the above table.

Note: This table should only be used when the sample mean and the sample standard deviation are calculated from the same set of data.

TABLE ST3 — The correlation test

Sample size	Confidence					
	90%	95%	98%	99%	99.8%	99.9%
4	.900	.950	.980	.990	.998	.999
5	.805	.878	.934	.959	.986	.991
6	.729	.811	.882	.917	.963	.974
7	.669	.754	.833	.875	.935	.951
8	.621	.707	.789	.834	.905	.925
9	.582	.666	.750	.798	.875	.898
10	.549	.632	.715	.765	.847	.872
11	.521	.602	.685	.735	.820	.847
12	.497	.576	.658	.708	.795	.823
13	.476	.553	.634	.684	.772	.801
14	.456	.532	.612	.661	.750	.780
15	.441	.514	.592	.641	.730	.760
16	.426	.497	.574	.623	.711	.742
17	.412	.482	.558	.606	.694	.725
18	.400	.468	.543	.590	.678	.708
19	.389	.456	.529	.575	.662	.693
20	.378	.444	.516	.561	.648	.679
25	.336	.396	.461	.506	.585	.613
30	.307	.363	.423	.464	.540	.570
40	.264	.313	.367	.400	.470	.497
60	.215	.254	.300	.331	.390	.413

Calculated value = |the sample correlation coefficient|.

TABLE ST4(a) — The SD test — 95% confidence

Degrees of freedom for smaller SD	Degrees of freedom for larger SD														
	1	2	3	4	5	6	7	8	9	10	12	15	20	60	Infinity
1	25.45	28.28	29.40	29.99	30.36	30.61	30.79	30.93	31.03	31.12	31.25	31.38	31.51	31.78	31.91
2	6.21	6.24	6.26	6.26	6.27	6.27	6.27	6.27	6.28	6.27	6.28	6.28	6.28	6.28	6.28
3	4.18	4.00	3.93	3.89	3.86	3.84	3.82	3.81	3.80	3.80	3.79	3.77	3.76	3.74	3.73
4	3.50	3.26	3.16	3.10	3.06	3.03	3.01	3.00	2.98	2.97	2.96	2.94	2.93	2.89	2.87
5	3.16	2.90	2.79	2.72	2.67	2.64	2.62	2.60	2.58	2.57	2.55	2.54	2.52	2.47	2.45
6	2.97	2.69	2.57	2.50	2.45	2.41	2.39	2.37	2.35	2.34	2.32	2.30	2.27	2.23	2.20
7	2.84	2.56	2.43	2.35	2.30	2.26	2.23	2.21	2.20	2.18	2.16	2.14	2.11	2.06	2.03
8	2.75	2.46	2.33	2.25	2.20	2.16	2.13	2.10	2.09	2.07	2.05	2.02	2.00	1.94	1.92
9	2.69	2.39	2.25	2.17	2.12	2.08	2.05	2.02	2.01	1.99	1.97	1.94	1.92	1.86	1.82
10	2.63	2.34	2.20	2.11	2.06	2.02	1.99	1.96	1.94	1.93	1.90	1.88	1.85	1.79	1.75
12	2.56	2.26	2.11	2.03	1.97	1.93	1.90	1.87	1.85	1.84	1.81	1.78	1.75	1.69	1.65
15	2.49	2.18	2.04	1.95	1.89	1.85	1.81	1.79	1.77	1.75	1.72	1.69	1.66	1.59	1.55
20	2.42	2.11	1.96	1.87	1.81	1.77	1.73	1.71	1.69	1.66	1.64	1.60	1.57	1.49	1.45
60	2.30	1.98	1.83	1.73	1.67	1.62	1.58	1.55	1.53	1.51	1.47	1.44	1.39	1.29	1.27
Infinity	2.24	1.92	1.77	1.67	1.60	1.55	1.51	1.48	1.45	1.43	1.39	1.35	1.31	1.18	1.00

Calculated value = (larger SD)/(smaller SD)

TABLE ST4(b) — The SD test — 99% confidence

Degrees of freedom for smaller SD	Degrees of freedom for larger SD														
	1	2	3	4	5	6	7	8	9	10	12	15	20	60	Infinity
1	127.32	141.42	147.02	150.00	151.84	153.09	154.00	154.67	155.21	155.64	156.29	156.94	157.59	158.91	159.58
2	14.09	14.11	14.11	14.11	14.12	14.12	14.12	14.12	14.12	14.12	14.12	14.12	14.12	14.12	14.12
3	7.45	7.06	6.89	6.80	6.74	6.70	6.67	6.64	6.62	6.61	6.58	6.56	6.54	6.49	6.47
4	5.60	5.13	4.93	4.81	4.74	4.69	4.65	4.62	4.60	4.58	4.55	4.48	4.49	4.43	4.40
5	4.77	4.28	4.07	3.94	3.87	3.81	3.77	3.74	3.71	3.69	3.66	3.63	3.59	3.52	3.48
6	4.32	3.81	3.59	3.47	3.39	3.33	3.28	3.25	3.22	3.20	3.17	3.13	3.10	3.02	2.98
7	4.03	3.52	3.30	3.17	3.09	3.03	2.98	2.95	2.92	2.89	2.86	2.82	2.78	2.70	2.66
8	3.83	3.32	3.10	2.97	2.88	2.82	2.77	2.74	2.71	2.69	2.65	2.61	2.57	2.49	2.44
9	3.69	3.18	2.95	2.82	2.73	2.67	2.62	2.59	2.56	2.53	2.50	2.46	2.41	2.33	2.28
10	3.58	3.07	2.84	2.71	2.62	2.56	2.51	2.47	2.44	2.42	2.38	2.34	2.30	2.20	2.15
12	3.43	2.92	2.69	2.55	2.46	2.40	2.35	2.31	2.28	2.26	2.22	2.17	2.13	2.03	1.97
15	3.29	2.77	2.55	2.41	2.32	2.25	2.20	2.16	2.13	2.10	2.06	2.02	1.97	1.87	1.81
20	3.15	2.64	2.41	2.27	2.18	2.11	2.06	2.02	1.99	1.96	1.92	1.87	1.82	1.71	1.64
60	2.91	2.44	2.17	2.03	1.94	1.87	1.81	1.77	1.73	1.70	1.66	1.60	1.55	1.40	1.30
Infinity	2.81	2.30	2.07	1.93	1.83	1.76	1.70	1.66	1.62	1.59	1.54	1.48	1.41	1.24	1.00

Calculated value = (larger SD)/(smaller SD)

TABLE ST5 — Normal distribution percentages

Standardized value	% greater than the value	Standardized value	% greater than the value	Standardized value	% greater than the value	Standardized value	% greater than the value	Standardized value	% greater than the value	Standardized value	% greater than the value
0.00	50.0	0.40	34.5	0.80	21.2	1.20	11.5	1.60	5.48	2.00	2.28
0.01	49.6	0.41	34.1	0.81	20.9	1.21	11.3	1.61	5.37	2.05	2.02
0.02	49.2	0.42	33.7	0.82	20.6	1.22	11.1	1.62	5.26	2.10	1.79
0.03	48.8	0.43	33.4	0.83	20.3	1.23	10.9	1.63	5.16	2.15	1.58
0.04	48.4	0.44	33.0	0.84	20.1	1.24	10.8	1.64	5.05	2.20	1.39
0.05	48.0	0.45	32.6	0.85	19.8	1.25	10.6	1.65	4.95	2.25	1.22
0.06	47.6	0.46	32.3	0.86	19.5	1.26	10.4	1.66	4.85	2.30	1.07
0.07	47.2	0.47	31.9	0.87	19.2	1.27	10.2	1.67	4.75	2.35	0.94
0.08	46.8	0.48	31.6	0.88	18.9	1.28	10.0	1.68	4.65	2.40	0.82
0.09	46.4	0.49	31.2	0.89	18.7	1.29	9.9	1.69	4.55	2.45	0.71
0.10	46.0	0.50	30.9	0.90	18.4	1.30	9.7	1.70	4.46	2.50	0.62
0.11	45.6	0.51	30.5	0.91	18.1	1.31	9.5	1.71	4.36	2.55	0.54
0.12	45.2	0.52	30.2	0.92	17.9	1.32	9.3	1.72	4.27	2.60	0.47
0.13	44.8	0.53	29.8	0.93	17.6	1.33	9.2	1.73	4.18	2.65	0.40
0.14	44.4	0.54	29.5	0.94	17.4	1.34	9.0	1.74	4.09	2.70	0.35
0.15	44.0	0.55	29.1	0.95	17.1	1.35	8.9	1.75	4.01	2.75	0.30
0.16	43.6	0.56	28.8	0.96	16.9	1.36	8.7	1.76	3.92	2.80	0.26
0.17	43.3	0.57	28.4	0.97	16.6	1.37	8.5	1.77	3.84	2.85	0.22
0.18	42.9	0.58	28.1	0.98	16.4	1.38	8.4	1.78	3.75	2.90	0.19
0.19	42.5	0.59	27.8	0.99	16.1	1.39	8.2	1.79	3.67	2.95	0.16
0.20	42.1	0.60	27.4	1.00	15.9	1.40	8.1	1.80	3.59	3.00	0.14
0.21	41.7	0.61	27.1	1.01	15.6	1.41	7.9	1.81	3.51	3.05	0.11
0.22	41.3	0.62	26.8	1.02	15.4	1.42	7.8	1.82	3.44	3.10	0.097
0.23	40.9	0.63	26.4	1.03	15.2	1.43	7.6	1.83	3.36	3.15	0.082
0.24	40.5	0.64	26.1	1.04	14.9	1.44	7.5	1.84	3.29	3.20	0.069
0.25	40.1	0.65	25.8	1.05	14.7	1.45	7.4	1.85	3.22	3.25	0.058
0.26	39.7	0.66	25.5	1.06	14.5	1.46	7.2	1.86	3.14	3.30	0.048
0.27	39.4	0.67	25.2	1.07	14.2	1.47	7.1	1.87	3.07	3.35	0.040
0.28	39.0	0.68	24.8	1.08	14.1	1.48	6.9	1.88	3.01	3.40	0.034
0.29	38.6	0.69	24.5	1.09	13.8	1.49	6.8	1.89	2.94	3.45	0.028
0.30	38.2	0.70	24.2	1.10	13.6	1.50	6.7	1.90	2.87	3.50	0.023
0.31	37.8	0.71	23.9	1.11	13.4	1.51	6.6	1.91	2.81	3.55	0.019
0.32	37.5	0.72	23.6	1.12	13.1	1.52	6.4	1.92	2.74	3.60	0.016
0.33	37.1	0.73	23.3	1.13	12.9	1.53	6.3	1.93	2.68	3.65	0.013
0.34	36.7	0.74	23.0	1.14	12.7	1.54	6.2	1.94	2.62	3.70	0.011
0.35	36.3	0.75	22.7	1.15	12.5	1.55	6.1	1.95	2.56	3.75	0.009
0.36	35.9	0.76	22.4	1.16	12.3	1.56	5.9	1.96	5.50	3.80	0.007
0.37	35.6	0.77	22.1	1.17	12.1	1.57	5.8	1.97	2.44	3.85	0.006
0.38	35.2	0.78	21.8	1.18	11.9	1.58	5.7	1.98	2.39	3.90	0.005
0.39	34.8	0.79	21.5	1.19	11.7	1.59	5.6	1.99	2.33	3.95	0.004

$$\text{Standized value} = \frac{\text{value} - \text{mean}}{\text{standard deviation}}$$

Value = mean + (standardized value)(standard deviation)

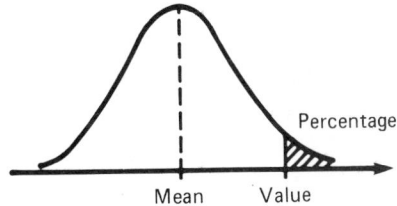

Percentage

Mean   Value

TABLE ST6 — Estimated population percentages

Item number	Sample size									
	1	2	3	4	5	6	7	8	9	10
1	50.0	29.3	20.6	15.9	12.9	10.9	9.4	8.3	7.4	6.7
2		70.7	50.0	38.6	31.5	26.6	23.0	20.2	18.1	16.3
3			79.4	61.4	50.0	42.2	36.5	32.1	28.7	25.9
4				84.1	68.5	57.8	50.0	44.0	39.4	35.6
5					87.1	73.5	63.5	56.0	50.0	45.2
6						89.1	77.1	67.9	60.7	54.8
7							90.6	79.8	71.3	64.4
8								91.7	81.9	74.1
9									92.6	83.7
10										93.3

Item number	Sample size									
	11	12	13	14	15	16	17	18	19	20
1	6.1	5.6	5.2	4.8	4.5	4.2	4.0	3.8	3.6	3.4
2	14.9	13.7	12.7	11.8	11.0	10.3	9.8	9.2	8.7	8.4
3	23.7	21.8	20.1	18.7	17.5	16.4	15.5	14.7	13.9	13.2
4	32.4	29.8	27.6	25.7	24.0	22.5	21.3	20.1	19.1	18.1
5	41.2	37.9	35.1	32.6	30.5	28.7	27.0	25.5	24.2	23.6
6	50.0	46.0	42.5	39.6	37.0	34.8	32.8	31.0	29.4	27.9
7	58.8	54.0	50.0	46.5	43.5	40.9	38.5	36.4	34.5	32.8
8	67.6	62.1	57.5	53.5	50.0	47.0	44.3	41.8	39.7	37.7
9	76.3	70.2	64.9	60.4	56.5	53.1	50.0	47.3	44.8	42.6
10	85.1	78.3	72.4	67.4	63.0	59.2	55.8	52.7	50.0	47.6
11	93.9	86.3	79.9	74.3	69.5	65.3	61.5	58.2	55.2	52.5
12		94.4	87.3	81.3	76.0	71.4	67.3	63.6	60.3	57.4
13			94.8	88.2	82.5	77.5	73.0	69.0	65.5	62.3
14				95.2	89.0	83.6	78.8	74.5	70.6	67.2
15					95.5	89.7	84.5	79.9	75.8	72.1
16						95.8	90.3	85.4	81.0	77.0
17							96.0	90.8	86.1	81.9
18								96.2	91.3	86.8
19									96.4	91.7
20										96.6

For sample sizes greater than 20.
Estimated % in population. = 100 (item number − 0.3)/(sample size + 0.4)

TABLE ST7 — Dixon's test

Sample size	Confidence level			
	90%	95%	98%	99%
4	0.765	0.831	0.889	0.922
5	0.642	0.717	0.780	0.831
6	0.560	0.621	0.698	0.737
7	0.507	0.570	0.637	0.694
8	0.468	0.524	0.590	0.638
9	0.437	0.492	0.555	0.594
10	0.412	0.464	0.527	0.564
12	0.376	0.429	0.482	0.520
14	0.349	0.397	0.450	0.485
16	0.329	0.376	0.426	0.461
18	0.313	0.354	0.407	0.438
20	0.300	0.340	0.391	0.417
25	0.277	0.316	0.362	0.386
30	0.260	0.300	0.341	0.368

The calculated value is equal to the larger of A or B

$$A = \frac{\text{(largest value)} - \text{(second largest value)}}{\text{range of the data}}$$

$$B = \frac{\text{(second smallest value)} - \text{(smallest value)}}{\text{range of the data}}$$

TABLE ST8 — The YES–NO test

Sample level	Confidence level					
	90%	95%	98%	99%	99.8%	99.9%
5	4½	—	—	—	—	—
6	5½	—	—	—	—	—
7	6½	6½	6½		—	—
8	6½	7½	7½	7½	—	—
9	7½	7½	8½	8½	—	—
10	8½	8½	9½	9½	9½	—
11	8½	9½	9½	10½	10½	10½
12	9½	9½	10½	10½	11½	11½
13	9½	10½	11½	11½	12½	12½
14	10½	11½	11½	12½	12½	13½
15	11½	11½	12½	12½	13½	13½
16	11½	12½	13½	13½	14½	14½
17	12½	12½	13½	14½	15½	15½
18	12½	13½	14½	14½	15½	15½
19	13½	14½	14½	15½	16½	16½
20	14½	14½	15½	16½	17½	17½
25	17½	17½	18½	19½	20½	20½
30	19½	20½	21½	22½	23½	24½
35	22½	23½	24½	25½	26½	27½
40	25½	26½	27½	28½	30½	30½
45	28½	29½	30½	31½	33½	33½
50	31½	32½	33½	34½	36½	36½
60	36½	38½	39½	40½	42½	43½
70	42½	43½	45½	46½	48½	49½
80	47½	49½	50½	51½	54½	55½
90	53½	54½	56½	57½	60½	60½
100	58½	60½	61½	63½	66½	66½

Calculated value = the number of 'yes's or the number of 'no's, which ever is the greater.

TABLE ST9 — Wilsoxon's matched pairs test

Sample size	Confidence level 90%	95%	98%	99%
5	$14\frac{1}{2}$	—	—	—
6	$18\frac{1}{2}$	$20\frac{1}{2}$	—	—
7	$24\frac{1}{2}$	$25\frac{1}{2}$	$27\frac{1}{2}$	—
8	$30\frac{1}{2}$	$32\frac{1}{2}$	$34\frac{1}{2}$	$35\frac{1}{2}$
9	$36\frac{1}{2}$	$39\frac{1}{2}$	$41\frac{1}{2}$	$43\frac{1}{2}$
10	$44\frac{1}{2}$	$46\frac{1}{2}$	$49\frac{1}{2}$	$51\frac{1}{2}$
11	$52\frac{1}{2}$	$55\frac{1}{2}$	$58\frac{1}{2}$	$60\frac{1}{2}$
12	$60\frac{1}{2}$	$64\frac{1}{2}$	$68\frac{1}{2}$	$70\frac{1}{2}$
13	$69\frac{1}{2}$	$73\frac{1}{2}$	$78\frac{1}{2}$	$81\frac{1}{2}$
14	$79\frac{1}{2}$	$83\frac{1}{2}$	$89\frac{1}{2}$	$92\frac{1}{2}$
15	$89\frac{1}{2}$	$94\frac{1}{2}$	$100\frac{1}{2}$	$104\frac{1}{2}$
16	$100\frac{1}{2}$	$106\frac{1}{2}$	$112\frac{1}{2}$	$116\frac{1}{2}$
17	$111\frac{1}{2}$	$118\frac{1}{2}$	$125\frac{1}{2}$	$129\frac{1}{2}$
18	$123\frac{1}{2}$	$130\frac{1}{2}$	$138\frac{1}{2}$	$143\frac{1}{2}$
19	$136\frac{1}{2}$	$143\frac{1}{2}$	$152\frac{1}{2}$	$157\frac{1}{2}$
20	$149\frac{1}{2}$	$157\frac{1}{2}$	$166\frac{1}{2}$	$172\frac{1}{2}$

Calculated value = the larger of the two rank totals.

TABLE ST10 — Wilcoxon's rank sum test

| Sample size | | Confidence level | | | |
Small	Large	90%	95%	98%	99%
3	3	14½	—	—	—
3	4	17½	—	—	—
3	5	19½	20½	—	—
3	6	21½	22½	—	—
3	7	24½	25½	26½	—
3	8	26½	27½	29½	—
3	9	28½	30½	31½	32½
3	10	31½	32½	34½	35½
4	4	24½	25½	—	—
4	5	27½	28½	29½	—
4	6	30½	31½	32½	33½
4	7	33½	34½	36½	37½
4	8	36½	37½	39½	40½
4	9	39½	41½	42½	44½
4	10	42½	44½	46½	47½
5	5	35½	37½	38½	39½
5	6	39½	41½	42½	43½
5	7	43½	44½	46½	48½
5	8	46½	48½	50½	52½
5	9	50½	52½	54½	56½
5	10	53½	56½	58½	60½
6	6	49½	51½	53½	54½
6	7	54½	56½	58½	59½
6	8	58½	60½	62½	64½
6	9	62½	64½	67½	69½
6	10	66½	69½	72½	74½
7	7	65½	68½	70½	72½
7	8	70½	73½	76½	77½
7	9	75½	78½	81½	83½
7	10	80½	83½	86½	88½
8	8	84½	86½	90½	92½
8	9	89½	92½	96½	98½
8	10	95½	98½	102½	104½
9	9	104½	108½	111½	114½
9	10	110½	114½	118½	121½
10	10	127½	131½	135½	138½

Calculated value = the larger of the two rank totals.

TABLE ST11 — The χ-squared test

Degrees of freedom	Confidence level		
	95%	99%	99.9%
1	3.84	6.64	10.83
2	5.99	9.21	13.82
3	7.82	11.35	16.27
4	9.49	13.28	18.47
5	11.07	15.08	20.51
6	12.59	16.81	22.46
7	14.07	18.49	24.36
8	15.51	20.09	26.31
9	16.92	21.67	27.89
10	18.31	23.21	29.59
11	19.68	24.72	31.26
12	21.03	26.22	32.91
13	22.36	27.69	34.51
14	23.68	29.14	36.12
15	25.00	30.58	37.70
16	26.30	32.00	39.25
17	27.59	33.41	40.79
18	28.87	34.81	42.31
19	30.14	36.19	43.82
20	31.41	37.57	45.32

Calculated value $= \sum \left[\dfrac{(O - E)^2}{E} \right]$

where O = observed frequency and E = expected frequency.

TABLE ST12 — Goldsmith's cusum test

Length of span	Confidence level 90%	95%	99%
5	2.4	2.7	3.3
6	2.7	3.0	3.6
7	2.9	3.2	4.0
8	3.2	3.5	4.3
9	3.4	3.7	4.6
10	3.6	3.9	4.9
11	3.8	4.1	5.1
12	3.9	4.3	5.3
13	4.0	4.5	5.5
14	4.1	4.6	5.6
15	4.2	4.8	5.8
20	5.2	5.6	6.8
25	5.6	6.0	7.3
30	6.2	6.7	8.0
40	7.2	7.8	9.3
50	8.0	8.6	10.4
60	8.8	9.5	11.3
70	9.5	10.3	12.2
80	10.1	10.8	12.9
90	10.5	11.3	13.6
100	11.0	11.8	14.3

$$\text{Calculated value} = \frac{|\text{maximum cusum}|}{\text{localized SD}}$$

The required values in the above table were obtained from a monogram in BS 5703, part 2.

Further reading

If you wish to investigate more advanced or more specialized techniques of data analysis you could turn to the later texts in this series. They will be written in the same style and format as this book. Alternatively you could explore the many thousands of books with the word 'statistics' in their titles. Unfortunately many statistics books are far too mathematical for the average industrial scientist so you could become very frustrated trying to find a suitable text by selecting at random. The following list contains books which are less mathematical and focus on practical applications, though they do vary in difficulty.

Barnett & Lewis, V. (1979) *Outliers in Statistical Data*, John Wiley, Chichester.

British Standards Institution (1975) BS 2846, *Guide to Statistical Interpretation of Data*, Parts I–VII, London.

British Standards Institution (1980) BS 5703, *Data Analysis and Quality Control using Cusum Techniques*, Parts I–IV, London.

Box, G. E. P., Hunter, W. G. & Hunter, J. S. (1978) *Statistics for Experimenters*, John Wiley, Chichester.

Caulcutt, R. (1983) *Statistics in Research Development*, Chapman & Hall, London.

Caulcutt, R. & Boddy, R. (1983) *Statistics for Analytical Chemists*, Chapman & Hall, London.

Daniel, C. (1976) *Applications of Statistics in Industrial Experimentation*, John Wiley, Chichester.

Davies, O. L. (1978) *Design and Analysis of Industrial Experiments*, Longman, Harlow.

Davies, O. L. & Goldsmith, P. L. (1972) *Statistical Methods in Research and Production*, Longman, Harlow.

Duncan, A. J. (1974) *Quality Control and Industrial Statistics*, Irwin, Homewood, Illinois.

Miller, J. C. & Miller, J. N. (1988) *Statistics for Analytical Chemistry*, Ellis Horwood, Chichester.

Neave, H. R. (1979) *Quick and Simple Tests Based on Extreme Observations*, Journal of Quality Technology, Vol. II, No. 2, pp. 66–79.

Oakland, J. S. (1986) *Statistical Process Control*, Heinemann, London.

Price, F. (1984) *Right First Time: Using Quality Control for Profit*, Gower, London.

Woodward, R. H. & Goldsmith (1964) Cumulative Sum Techniques, Oliver & Boyd, Edinburgh.

Index